AUTOMATION
for
FOOD
ENGINEERING

Food Quality Quantization
and Process Control

CRC Series in
CONTEMPORARY FOOD SCIENCE

Fergus M. Clydesdale, Series Editor
University of Massachusetts, Amherst

Published Titles:

America's Foods Health Messages and Claims:
Scientific, Regulatory, and Legal Issues
James E. Tillotson

New Food Product Development: From Concept to Marketplace
Gordon W. Fuller

Food Properties Handbook
Shafiur Rahman

Aseptic Processing and Packaging of Foods:
Food Industry Perspectives
Jarius David, V. R. Carlson, and Ralph Graves

The Food Chemistry Laboratory: A Manual for Experimental Foods,
Dietetics, and Food Scientists
Connie Weaver

Handbook of Food Spoilage Yeasts
Tibor Deak and Larry R. Beauchat

Food Emulsions: Principles, Practice, and Techniques
David Julian McClements

Getting the Most Out of Your Consultant: A Guide
to Selection Through Implementation
Gordon W. Fuller

Antioxidant Status, Diet, Nutrition, and Health
Andreas M. Papas

Food Shelf Life Stability
N.A. Michael Eskin and David S. Robinson

Bread Staling
Pavinee Chinachoti and Yael Vodovotz

Interdisciplinary Food Safety Research
Neal M. Hooker and Elsa A. Murano

Automation for Food Engineering: Food Quality Quantization
and Process Control
Yanbo Huang, A. Dale Whittaker, and Ronald E. Lacey

CRC Series in
CONTEMPORARY FOOD SCIENCE

AUTOMATION
for
FOOD
ENGINEERING

Food Quality Quantization
and Process Control

Yanbo Huang, Ph.D.
A. Dale Whittaker, Ph.D.
Ronald E. Lacey, Ph.D.

CRC Press
Boca Raton   London   New York   Washington, D.C.

Library of Congress Cataloging-in-Publication Data

Huang, Yanbo, 1958–
   Automation for food engineering : food quality quantization and process control / by
Yanbo Huang, Dale Whittaker, Ronald E. Lacey.
      p.   cm. — (CRC series in contemporary food science)
Includes bibliographical references and index.
   ISBN 0-8493-2230-8 (alk. paper)
      1. Food industry and trade—Automation. 2. Food—Quality—Evaluation—Automation.
I. Whittaker, Dale. II. Lacey, Ronald E. III. Title. IV. Series.

TP372.8 .H83 2001
664'.024—dc21   2001025618

Visit the CRC Press Web site at www.crcpress.com

Dedication

To Meixi, Tian, and Mamie, and my father, mother, and sister

Yanbo Huang

To my family

A. Dale Whittaker

I would like to dedicate this to my parents who believed in the power of education and to my wife, Sally, and children, Kathryn, Charlotte, and David, who do not always understand what I do or why I do it, but support me anyway

Ronald E. Lacey

Preface

Food quality quantization and process control are two important fields in the automation of food engineering. Food quality quantization is a key technique in automating food quality evaluation. Food quality process control is the focus in food production lines. In the past 10 years, electronics and computer technologies have significantly pushed forward the progress of automation in the food industry. Research, development, and applications of computerized food quality evaluation and process control have been accomplished time after time. This is changing the traditional food industry. The growth of applications of electronics and computer technologies to automation for food engineering in the food industry will produce more nutritious, better quality, and safer items for consumers.

The book describes the concepts, methods, and theories of data acquisition, data analysis, modeling, classification and prediction, and control as they pertain to food quality quantization and process control. The book emphasizes the applications of advanced methods, such as wavelet analysis and artificial neural networks, to automated food quality evaluation and process control and introduces novel system prototypes such as machine vision, elastography, and the electronic nose for food quality measurement, analysis, and prediction. This book also provides examples to explain real-world applications.

Although we expect readers to have a certain level of mathematical background, we have simplified this requirement as much as possible to limit the difficulties for all readers from undergraduate students, researchers, and engineers to management personnel. We hope that the readers will benefit from this work.

Outline of the Book

Six chapters follow the Introduction.

Chapter 2 concerns data acquisition (DAQ) from the measurement of food samples. In Chapter 2, the issues of sampling are discussed with examples of sampling for beef grading, food odor measurement, and meat quality evaluation. Then, the general concepts and systems structure are introduced. The examples of ultrasonic A-mode signal acquisition for beef grading, electronic nose data acquisition for food odor measurement, and snack food frying data acquisition for process quality control are presented. Imaging systems, as they are applied more and more in the area of food quality characterization, are discussed in a separate section. Generic machine vision systems and medical imaging systems are described. Image acquisition for snack food quality evaluation, ultrasonic B-mode imaging for beef grading, and elastographic imaging for meat quality evaluation are presented as examples.

Chapter 3 is about processing and analysis of acquired data. In this chapter, the methods of data preprocessing, such as data scaling, Fourier transform, and wavelet transform are presented first. Then, the methods of static and dynamic data analysis are described. Examples of ultrasonic A-mode signal analysis for beef grading, electronic nose data analysis for food odor measurement, and dynamic data analysis of snack food frying process are presented. Image processing, including image preprocessing, image segmentation, and image feature extraction, is discussed separately. The methods of image morphological and textural feature extraction (such as Haralick's statistical and wavelet decomposition) are described. Examples of segmentation of elastograms for the detection of hard objects in packaged beef rations, morphological and Haralick's statistical textural feature extraction from images of snack food samples, Haralick's statistical textural and gray-level image intensity feature extraction from ultrasonic B-mode images for beef grading, and Haralick's statistical and wavelet textural feature extraction from meat elastograms are presented.

Chapter 4 concerns modeling for food quality quantization and process control. Model strategies, both theoretical and empirical, are discussed first in this chapter. The idea of an input–output model based on system identification is introduced. The methods of linear statistical modeling and ANN (artificial neural network) -based nonlinear modeling are described. In dynamic process modeling, the models of ARX (autoregressive with exogenous input) and NARX (nonlinear autoregressive with exogenous input) are emphasized. In statistical modeling, examples of modeling based on ultrasonic A-mode signals for beef grading, meat attribute prediction modeling based on Haralick's statistical textural features extracted from ultrasonic elastograms, and snack food frying process ARX modeling are presented. In ANN modeling, the examples of modeling for beef grading, modeling for food odor pattern recognition with electronic nose, meat attribute prediction modeling, and snack food frying process NARX modeling are presented.

Chapter 5 discusses classification and prediction of food quality. In this chapter, the methods of classification and prediction for food quality quantization are introduced first. Examples of beef sample classification for grading based on statistical and ANN modeling, electronic nose data classification for food odor pattern recognition, and meat attribute prediction based on statistical and ANN modeling are presented. For food quality process control, the methods of one-step-ahead and multiple-step-ahead predictions of linear and nonlinear dynamic models, ARX and NARX, are described. The examples of one-step-ahead and multiple-step-ahead predictions for the snack food frying process are presented.

Chapter 6 concentrates on food quality process control. In this chapter, the strategies of IMC (internal model control) and PDC (predictive control) are introduced. Based on the linear IMC and PDC, the ANN-based nonlinear IMC and PDC, that is, NNIMC (neural network-based internal model control) and NNPDC (neural network-based predictive control), are extended and described. The algorithms for controller design also are described. The methods of controller tuning are discussed. The examples of NNIMC and neuro-fuzzy PDC for the snack food frying process are presented.

Chapter 7 concludes the work. This chapter is concerned with systems integration for food quality quantization and process control. In this chapter, based on the discussion and description from the previous chapters concerning system components for food quality quantization and process control, the principles, methods, and tools of systems integration for food quality quantization and process control are presented and discussed. Then, the techniques of systems development, especially software development, are discussed for food quality quantization and process control.

Yanbo Huang

A. Dale Whittaker

Ronald E. Lacey
College Station, Texas
May 2001

Acknowledgments

Numerous individuals provided help in the process of drafting this book. We mention them to express our thanks. Our special thanks go to Ms. Sandy Nalepa, Dr. Stephen B. Smith, and Mr. Bill Hagen for their permission to use ASAE, Journal of Animal Science, and IEEE copyrighted materials. Our special thanks also go to Dr. Bosoon Park, Mr. David Bullock, Mr. Brian Thane, and Mr. Wei Wang for their permission to use content from their dissertations and theses.

We are very grateful for professional English editing by Ms. Judy Wadsworth. Our sincere appreciation goes to Ms. Lourdes M. Franco, former CRC editor, for her encouragement and valuable suggestions. Our sincere appreciation also goes to Ms. Carol Hollander, CRC senior editor, and Ms. Naomi Rosen, CRC editorial assistant, for their valuable help and suggestions for submission of the final manuscript. Finally, we thank the Department of Agricultural Engineering at Texas A&M University for offering a comfortable environment for our research and development.

About the Authors

Yanbo Huang is Associate Research Scientist at Texas A&M University. He earned his B.S. degree in Industrial Automation from Beijing University of Science and Technology, M.S. in Industrial Automation from the Chinese Academy of Mechanics and Electronics Sciences, and Ph.D. from Texas A&M University. Since 1991, he has been conducting research on food quality evaluation automation and process control at Texas A&M University.

A. Dale Whittaker is Professor Agricultural Engineering at Texas A&M University. He earned his B.S. degree from Texas A&M University, and M.S. and Ph.D. degrees from Purdue University. He was the Acting Director of the Institution of Food Science and Engineering and the Director of the Food Processing Center under the auspices of the Institute at Texas A&M University.

Ronald E. Lacey is Associate Professor of Agricultural Engineering at Texas A&M University. He earned his B.S., M.S., and Ph.D. degrees from the University of Kentucky. Before his academic career, Ron worked in the food industry for nine years doing engineering, R&D, and management.

Contents

chapter one

Introduction

1.1 Food quality: a primary concern of the food industry

The quality of foods is of primary importance to the food industry and to consumers. Consumers want food products that are good, nutritious, and safe. High quality food products can boost the profitability of the food supply chain from farming, processing, and production to sales, thus, strengthening the entire enterprise. However, any failure of a food product may result in a consumer returning the product to the seller, writing a complaint letter to the manufacturer, or even filing a lawsuit against the food company. The failure may be the under fill of a package, off-flavor, odor, staleness, discoloration, defective packaging, expired shelf life, incurred illness, and so on. For the sake of meeting consumers' needs, the food industry has the obligation to produce food items that are uniform in quality, nutritious, and safe. A food company needs to have adequate quality assurance systems and active quality control systems to keep its products competitive in the market.

1.2 Automated evaluation of food quality

Evaluation is the key to ensuring the quality of food products. Often, evaluation detects component adequacy and documents mechanical, chemical, and microbiological changes over the shelf life of food items. Both qualitative and quantitative evaluation can provide the basis for determining if a food product meets target specifications. This quality information also provides feedback for adjustments to processes and products needed to achieve target quality

There are two methods for evaluation of food quality. One is subjective, based on the judgment of human evaluators. The other is objective, based on observations excluding human evaluators' opinions.

Subjective methods require the human evaluators to give their opinions regarding the qualitative and quantitative values of the characteristics of the food items under study. These methods usually involve sensed perceptions of texture, flavor, odor, color, or touch. However, even though the evaluators are highly trained, their opinions may vary because of the individual variability involved. Sensory panels are a traditional way to evaluate food quality. Although highly trained human evaluators are intelligent and able to perceive deviation from food quality standards, their judgments may not be consistent because of fatigue or other unavoidable mental and physical stresses.

The output of food quality evaluation is the primary basis for establishing the economic value of the food products for farmers, manufacturers, and consumers, and it can be useful for quality control of food products. Because traditional manual quality control is time-consuming, can be error prone, and cannot be conducted in real time, it has been highly desirable for the food industry to develop objective methods of quality evaluation for different food products in a consistent and cost-effective manner. The objective methods of food quality evaluation are based on scientific tests rather than the perceptions of human evaluators. They can be divided into two groups:

1. Physical measurement methods are concerned with such attributes of food product quality as size, texture, color, consistency, and imperfections. There are several sensors adapted for the physical evaluation of food product quality.
2. Chemical measurement methods test for enzyme, moisture, fiber, pH, and acidity. In many cases, these tests can be used to determine nutritive values and quality levels.

The development of computer and electronics technologies provides strong support to fast, consistent signal measurement, data collection, and information analysis. The greatest advantage of using computer technology is that once the food quality evaluation systems are set up and implemented, the system will perceive deviation from food quality standards in a consistent way and not experience the mental and physical problems of human evaluators. Another major benefit of using computer technology in food quality systems is that it is possible to integrate a large number of components to automate the processes of food quality evaluation. This automation can result in objective, fast, consistent food quality evaluation systems, a significant advancement for food engineering and industry.

This book will focus on the techniques for objective and automated food quality evaluation, especially nonintrusive/noninvasive food quality evaluation.

1.3   Food quality quantization and process control

Food quality quantization allows information to be represented numerically in a mathematical expression. The process of the representation is often automated. In evaluation, indicators of food quality such as analytical, on-line sensor, and mechanical measurements of food samples need to be quantized in use for assessing quality. Basically, food quality quantization is the mimic of human intelligence with machine intelligence in food quality evaluation. The machine intelligence is used to "view," "touch," and/or "taste" the food samples and, then, to differentiate them in a way that is often guided by results from a human sensory panel. The performance of the quantization is usually measured by a comparison of the quantitative data with sensory, classification assignment, or mechanical and chemical attribute measurements.

In general, the procedure of food quality quantization is as follows:

1. Sampling—the first step in food quality quantization involves collecting food samples according to a designed sampling procedure. The procedure is designed to produce enough data samples under different experimental conditions to be able to draw a conclusion with a certain statistical significance. When the samples are extracted, they need to be further processed, stored, and delivered onto the experimental board for measurement.
2. Data acquisition—sensors and transducers measure the collected food samples. The electrical signals indicate physical properties of the food products. The signal data are conditioned, converted, and stored for later processing and analysis.
3. Data processing and analysis—the data are processed, usually scaled or normalized, to produce a consistent magnitude between variables. The relationships between variables are tested and correlations between variables are determined. This step helps make decisions based on modeling strategy.
4. Modeling—mathematical models are statistically built between the (input) variables of the physical property measurements of food samples and the (output) variables of human sensory quantities, classification assignments, or mechanical and chemical measurements of the samples. The models determine quantitative relationships between the input and output variables.
5. Classification or prediction—based on the models, the food samples can be classified or predicted for their sensory, mechanical, and chemical attributes. The accuracy of the classification or prediction is calculated.

In this way the performance of the quantization can be evaluated based on the accuracy of the classification or prediction. If the performance is satisfactory, the food quality quantization scheme can be used in practical food quality evaluation; otherwise, it becomes necessary to reassess the modeling, data processing and analysis, data acquisition, or even sampling procedures to locate the spot(s) to refine the scheme. Figure 1.1 shows a diagram of the procedure for food quality quantization.

Food quality process control occurs when the difference between measurements of actual food quality and specifications of food quality is used to adjust the variables that can be manipulated to change product quality. The variables that indicate food quality may be quantities like color and

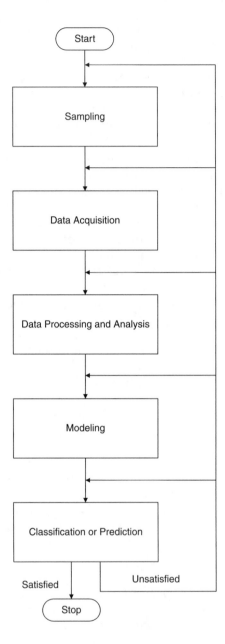

Figure 1.1 Diagram of the procedure for food quality quantization.

moisture content. The adjusted variables are quantities such as inlet temperature and material cooling conveyor speed in a continuous snack food
frying process. These adjustments are made based on the computation of
certain algorithms in an attempt to eliminate the deviation of observed
quality from the target specifications for the product. In general, the adjust-

ments can be done either when the problems occur or on a regular basis. The former is quality control (Besterfield, 1990) while the latter is process control, the topic of this book. Food quality process control applies methods and tools of process control to adjust the operating conditions in terms of the quality specifications of food processes for the consistency of food product quality.

In general, the procedure of food quality process control is as follows:

1. Sampling—sampling is performed in terms of the requirements of food quality process control. It needs to design experiments which produce enough data samples in different conditions with certain statistical significance. For effective process control, the collected samples need to be able to produce the measured data to cover the designated frequency range to represent the process dynamics sufficiently.
2. Data acquisition—with the prepared samples, the values of the food quality indication variables are measured by sensors and transducers, and the corresponding data for process operating conditions are recorded.
3. Data processing and dynamic analysis—the data are processed, usually scaled or normalized, to produce a consistent magnitude between variables. The dynamic relationships between variables are tested. The autocorrelations and cross correlations between variables are determined. This step helps make the decision about the process modeling strategy.
4. Modeling—linear or nonlinear discrete-time dynamic mathematical models are statistically built between the (input) variables of the levels of actuators in food processes and the (output) variables of food quality indication. The models determine quantitative, dynamic relationships between the input and output variables.
5. Prediction—based on the models, the quantities of food quality indications can be predicted in one-step-ahead or multiple-step-ahead modes. The accuracy of the predictions reflects the capability of the prediction models in control loops.
6. Controller design—the built process models are used to design the controllers based on certain algorithms. The controllers are tuned in order to perform well in the regulation of the process operating conditions to ensure the consistent quality of the final products.

The performance of the process control systems can be evaluated based on the specifications and requirements of the food processes. If the performance is satisfactory, the food quality process control scheme may be implemented in practice; otherwise, it needs to go back to the starting points of controller design, prediction, modeling, data processing and dynamic analysis, data acquisition, or even an experiment design to locate places to refine the scheme. Figure 1.2 shows the diagram of the procedure for food quality process control.

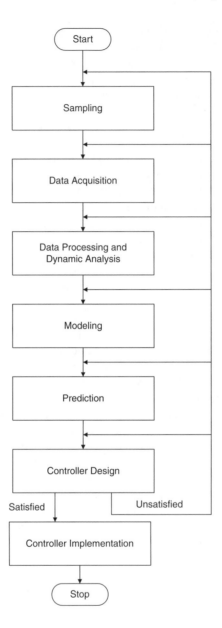

Figure 1.2 Diagram of the procedure for food quality process control.

The food quality quantization is the key technique for objective, automated food quality evaluation. It provides a quantitative representation of
food quality. Food quality process control uses the feedback from food
quality evaluation to make adjustment in order to meet food quality specifications in the process of food production. This book will focus on technical aspects of food quality quantization for establishment of objective,

automated food quality evaluation systems and food quality process control in the production of food products for consistent quality in food processes.

1.4   *Typical problems in food quality evaluation and process control*

Various problems need to be solved in food quality evaluation and process control. Over the past 10 years, we have conducted research and development projects at different size operations in this area. These projects have resolved some typical problems in food quality evaluation and process control. Throughout this book, a number of our projects in food quality evaluation and process control will be presented in chapter by chapter examples for the purpose of explanation of the technical aspects from sampling, data acquisition, data processing and analysis, modeling, classification and prediction, to control. By going though the problem solving processes in these projects, related concepts, methods, and strategies can be understood better. Next, the basis of three typical projects in our research and development work will be initially described by stating the problems which need to be solved through food quality quantization and process control. These projects will be further described in subsequent chapters in terms of the topics, and some other projects also will be involved for the purpose of explanation.

1.4.1   *Beef quality evaluation*

The U.S. Standards for Grades of Carcass Beef was revised a number of times from its tentative version in 1916 to the current version (USDA, 1997). In these standards, the quality grading system is designed to segregate beef carcasses into groups based on expected difference in quality using the subjective evaluation of human graders. When assigning quality grade scores, highly trained graders evaluate the degree of marbling in the lean surface at the 12th to 13th rib interface and the overall maturity of the carcass. This grading system is considered to be the best system available for determining expected beef quality. However, owing to the subjectivity involved, variations exist between graders although they are highly trained for the task. The grading process is primarily based on establishing the value of the carcass to the ranchers, packers, and consumers. The formulation of a methodology with alternative techniques is desired to evaluate objectively the carcass quality for the purpose of grading. The methodology is expected to be more accurate and consistent to better segregate products and, thereby, to reduce the variation in products.

In industry operations, with the growth of the beef industry, the speed of lines has increased dramatically. The subjective, inconsistent, inaccurate, and slow grading of live and slaughtered beef cattle combined with the wide variations in beef cattle genetics and management may cause wide variations in beef products. In order to reduce and even avoid such wide variations in beef products, methods need to be developed for fast, objective, consistent, and accurate determination of quality and yield grades of beef carcasses and the rank of live beef animals.

Computerized electronic instrumentation was considered as an alternative for fast and objective beef grading. This kind of grading system would objectively measure, classify, and predict the quality of beef cattle in order to have a beef product with high quality consistency. In later chapters, the projects of applications of ultrasound-based methods will be introduced in developing objective, noninvasive, nondestructive, and rapid beef quality evaluation systems.

1.4.2   Food odor measurement

In characterizing food quality, odor can be a useful parameter. Humans tend to utilize odor as an indicator of quality. Pleasant odors can enhance overall quality of the interaction with a particular system, whether it is the aroma of fresh coffee or baking bread. Unpleasant odors act as a signal that there is a problem, for example, spoilage in a food product. Unfortunately, odors are difficult to measure because they are usually comprised of a complex mixture of volatile molecules in small (i.e., parts per billion) concentrations.

Typically, odor measurement depends on human sensory evaluation (olfactometry) to establish odor parameters. Odors can be classified on a quantitative basis (odor intensity) or on a qualitative basis (odor hedonics). However, olfactometry has limitations in establishing specific odor measurements that fully characterize a system of interest. Hence, there have been ongoing efforts to develop a biomimicry instrument (a.k.a. the electronic nose) that can replace or supplement olfactory measurements. For example, the intensity and offensiveness of objectionable odors is of interest to food consumers. Continuous or semicontinuous measurements of an odor are not practical using olfactometry because of cost and methodology constraints. An electronic nose could provide continuous measurements in the same manner as a weather station provides ongoing measurements of wind speed, direction, and precipitation. Electronic noses cannot completely mimic the human nose. Considerations in applying an electronic nose to specific food products must be addressed.

Technology is needed for the quantization of aromas. In the upcoming chapters, the work of the electronic nose will be introduced in developing the technique of "smelling" the odors of food products, for example, detecting high temperature curing off-flavors in peanuts (Osborn et al., in press), that can measure the quality of foods continuously, fast, objectively, and nondestructively.

1.4.3   Continuous snack food frying quality process control

There are two basic types of frying for snack food items: batch and continuous. Batch frying is typically used in small-scale operations, such as restaurants. Continuous frying is used at large-scale operations such as the snack food industry. In continuous frying, a continuous input of snack food

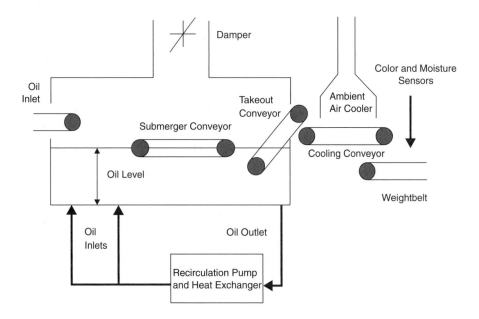

Figure 1.3 Diagram of the continuous snack food frying system.

material is put in one end of a fryer, pushed through by a submerger and oil flow, and, then, extracted at the other end.

The fryer for snack chips can be sketched as shown in Figure 1.3. The raw material exists in an extruder in the form of a ribbon. Along the inlet belt, a cutter slices the ribbon into set lengths. The cut ribbon is dropped into the oil of the fryer where it encounters a short free-float zone. A submerger that covers most of the length of the fryer pushes the chips below the surface of the oil and carries them to the other end. After the submerger, the chips are taken out by a conveyor. Then, the chips are placed on a separate cooling conveyor where ambient air flow cools them. Finally, the chips are transferred to a weight belt where sensors measure the brightness (or color intensity) and moisture content, indicators of the final product quality. The quality control is maintained by monitoring and stabilizing the levels of these indicators in terms of setpoints of the process operation. If human operators are involved in the control, they may be subject to overcorrection and overreaction to normal process variability.

In order to ensure consistency in product quality, automatic control is desired. The snack food frying process is complex. It has complicated interactions between the product quality indication variables and those factors that can be changed independently to affect these product quality indication variables. For the purpose of controlling such a complex process, it is first necessary to model it. Owing to the complexity of the snack food frying process, it is difficult to develop an analytical model of it. In order to determine heat transfer to the chips and subsequent moisture loss, a multitude of parameters must be considered, such as

- Oil inlet and outlet temperature.
- Oil mass flow.
- Raw material flow rate.
- Material moisture content at the inlet and outlet.
- Chip oil content, and so on.

However, even though all of the preceding parameters have been considered, it is still difficult to model a fryer operation in this manner. Instead, it is more feasible to develop an input–output model using the idea of system identification which views the process as a "black box" based on the input–output process model to develop a strategy to control the operation of the process. Based on process models, the following chapters will present the manner in which system identification is used to model effectively the frying process and the manner in which the food quality controllers are developed.

1.5   How to learn the technologies

This book focuses on using computer and electronics technologies to develop rapid and objective methods for food quality evaluation and process control. Readers are strongly encouraged to relate their problems to the relevant parts or to the entire book. Readers can understand the related concepts and methods and, then, go over the corresponding examples in related chapters to learn from real-world practice. Readers can find the references cited in the book for more details and consult with the authors for suggestion of solutions.

This book intends to provide a technical direction for application development of technologies for food quality evaluation and process control. It supplies theoretical descriptions and practical examples. Readers can study them, get hints, and solve their own problems. Similar to other engineering fields, application development in food engineering is a science and, also, an art. It is a good practice to extract the nature from the precedent and, based on the study, form your own structure to solve the problem. In this way, you will work out a unique technique for the problem solution. This is significant in both theory and practice.

References

Besterfield, Dale H., *Quality Control.*, 3rd ed., Prentice-Hall, Englewood Cliffs, NJ, 1990.

Osborn, G. S., Lacey, R. E., and Singleton, J. A., A method to detect peanut off-flavors using an electronic nose, *Trans. ASAE,* in press.

U.S. Department of Agriculture, *Official United States Standards for Grades of Carcass Beef.,* 1997.

chapter two

Data acquisition

Data generation and collection is generally the first step in food quality evaluation and process control. The work of data generation and collection is often done through computer based data acquisition (DAQ). In the food industry, DAQ techniques are often combined with other tools or methods to measure unique quality related parameters such as texture, color, or aroma. The acquired data may be single quantities (scalar), one-dimensional sequences or time series (vector), or two-dimensional images (matrix) in representation of selected food parameters. In this chapter, concepts of DAQ for food quality are established through the description of the software and hardware. The acquisition of images will be discussed separately, because they are being applied more frequently in food engineering. In addition, real-world examples will be provided to help understand the concepts with practical DAQ systems for food quality.

2.1 Sampling

As a first step in data acquisition, a sampling scheme for the product must be devised. The number of product units produced in a food processing plant is typically so large that it precludes testing the entire population. Thus, sampling is the process used to select objects for analysis so that the population is represented by the samples.

Consequently, a sampling procedure must be designed to select samples from a population to estimate the quality characteristics of the population. In order to develop an automated system for food quality evaluation, sample data are first acquired instrumentally, then they are analyzed. Mathematical models of quality classification and prediction are applied and, finally, control systems are developed on the basis of the models. Obviously, the performance of the control system is closely related to the sample selection. Samples need to sufficiently represent the population to be considered. In other words, in order to infer the variations of the properties of a population, it is necessary to select enough representative samples from each section of the population.

The following items are important to consider in developing a sampling plan.

1. Clearly establish the goal of the project—what are the critical parameters related to the quality of the food product? For example, the tenderness of beef muscle may be the primary attribute that determines the acceptability of the product to the consumer. Thus, a measure of tenderness is the critical parameter in an automated quality evaluation method. Clearly establishing the goal leads directly to the second consideration.

2. Determine the quantitative measurements that best represent the critical parameters identified in the goal—the critical parameters of food quality are often not easily measured using quantitative methods because they are qualitative attributes (i.e., flavor, texture, aroma). However, in order to develop an automated control system, a representative quantitative measurement of that parameter must be devised. Continuing the example of beef tenderness, much research has gone into establishing instrumental methods for measuring tenderness including various stress–strain techniques, correlation of electrical properties, correlation of chemical properties, and image analysis for determining fat distribution (i.e., marbling).

    The quantitative measurements used also establish the data types to be evaluated. As previously noted, single point measurements are scalar quantities (one-dimension—the measurement), time series are vector quantities (two-dimensional—the measurement and the time variable), and images are matrix quantities (three-dimensions—the measurement, the x coordinate, and the y coordinate). The data type has implications for the post sampling data processing methods to be applied, the computational time required, and the storage space needed for intermediate and final calculation results. While the continuing trend toward faster computers and less expensive storage continues, large data sets are often generated through higher order data types (i.e., matrix vs. vector) and increased sampling rates.

    Finding a representative instrumental measurement is not always easy and obtaining good correlation with the quality parameter may not always be possible. However, there are numerous examples, some discussed in this text, of quantitative measurements that do represent qualitative properties of a food.

3. Establish the control limits—automated control systems monitor *control variables* (i.e., the quantitative measures established in step 2) and adjust the process through *manipulated variables* based on the deviation from some desired condition (i.e., the *setpoint*). The upper and lower boundaries of the control variables, beyond which the product is unacceptable, determine the number of samples and the frequency of sampling necessary to represent the quality parameters of the population.

Representative samples must be taken from the population to determine quality characteristics. Therefore, some prior assumptions must be made for the selection of samples. If the selected samples do not represent the underlying population, the measurements from these samples will not make any sense and, if these results are used, they will produce incorrect conclusions.

Sampling can become a very complicated issue. The following questions must be answered:

1. Where is the optimum location of the sampling point?
2. How should the samples be taken from the population?
3. With what frequency should the sample be selected?
4. How many samples are sufficient, that is, what is the optimum sample size?

Typically the sampling task is planned to select random samples from a lot to determine the acceptability of the quality level. Then, the samples are examined for one or more characteristics and the decision is made on acceptance or rejection of the lot.

The determination of the size of samples to be selected is important. The number of samples should accurately and adequately describe the population. There are two extremes in sampling: none of the samples are selected, that is, 0 sample size and all of the samples are selected, that is, 100 percent sample size. On some occasions, no samples at all need to be drawn because some materials in a process do not need to be examined. In general, 100 percent sampling is not efficient. Statistical sampling techniques work in the area between 0 sampling and 100 percent sampling.

Randomly taken samples are useful in representing the characteristics of the population because they are selected in such a way that every item in the population has an equal chance to be chosen. Basically, there are simple random sampling and clustered random sampling. If "good" or "bad" products scatter uniformly over a population, then use of simple random sampling is appropriate. In many cases the population is divided into small subgroups, such as lots, each of which has products with uniform quality distribution, in which case clustered random sampling needs to be used.

Details of techniques for sample selection and preparation are beyond the scope of this book. Interested readers can refer to textbooks on the topics of Sample Surveys, Experimental Design, and Quality Control.

2.1.1   Example: Sampling for beef grading

For a study to grade beef samples, Whittaker et al. (1992) had 57 slaughtered test animals and a subset of 35 live test animals. The test animals (Simmental–Angus crossbred steers) were all classified as "A" carcass maturity (9 to 30 months of age). Each slaughtered test animal was scheduled to be measured no later than 15 min after the animal was killed. By measuring slaughtered animals, the variability in each measurement owing to respiratory action, blood circulation, muscle contraction, and other dynamic body functions caused by animal body movements were expected to be reduced or eliminated. Measurement of live animals was limited to a subset of the total number of test animals because of the restrictions placed on these animals owing to the excessive preslaughter handling by other research projects being conducted at the time. The measurement of live animals was needed primarily to determine the degree of difference existing between the measurements of each test animal when measured alive and immediately after slaughter.

The process for this study was categorized into the following phases.

1. Sampling.
2. Measuring.
3. Data acquisition.
4. Data preprocessing.
5. Data processing.
6. Quality grading.
7. Data analysis.

In the sampling phase, each test animal was prepared by using an electric clipper to remove the hair from the measuring area, an area approximately 15 × 30 cm located over the 12th and 13th rib section on the upper back portion of the animal. This area was brushed clean and mineral or vegetable oil was applied to improve instrumental contact. The live animals were measured in squeeze chutes, and the slaughtered animals were measured on the abattoir floor while hanging from the rail by both hind legs.

Similarly, Park (1991) used fresh *longissimus dorsi* (LD) muscles for establishing an instrument method for beef grading. For this study, all animals had a carcass maturity of A (9 to 30 months of age). Trained graders classified the samples into different groups from Abundant to Practically Devoid. Steaks 30 mm thick were excised from the LD of the 13th rib from the right side of each carcass. A 75 mm × 42 mm × 30 mm sample was excised from each steak. In this way, 124 specimens were sampled from 124 beef carcasses ranging across the marbling classes.

The number of test animals measured for the study was determined by the number of head available for slaughter during the experimental process. The samples from 124 slaughtered test animals were planned to be measured in co-ordination with experimentation by the Animal Science Department of Texas A&M University.

The task of predetermining the exact number of test animals needed for this study was practically impossible because actual fat concentration in muscle could only be known through chemical analysis after each sample was measured instrumentally. Every effort, therefore, was made to sample the test animals expected to represent a uniform distribution based upon a prespecified number of test animals representing the USDA quality grades across each marbling category for A carcass maturity.

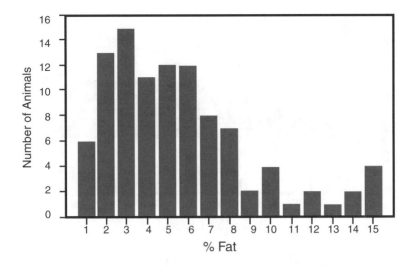

Figure 2.1. Profile of sample distribution. (From Park, 1991. With permission.)

Because of the limited number of test animals that were available for this study and the narrow distribution of fat classes within the marketplace, the distribution of test animals was not uniform. The sample number of most marbling categories was at least 6 specimens with the exception of sample categories located between 9 percent and 15 percent (very abundant marbling) as shown in Figure 2.1.

Thane (1992) had 90 live and 111 slaughtered beef animals available for instrumental measurement. Because of the accumulation of a high number of animals falling in the lower marbling classes (Traces, Slight, and Small), ranging in fat content from 2.5 percent to 5.0 percent, animals were randomly eliminated. Consequently, 71 live animals and 88 slaughtered carcasses remained for the study.

The Angus–Hereford crossbred heifers and Angus–Simmental crossbred steers were provided. The live weights of the Angus–Hereford crossbred heifers ranged from 338 to 585 kg (1290 lb), and their carcass weights ranged from 210 to 334 kg (737 lb). For the Angus–Simmental crossbred steers, live weights ranged from 332 to 683 kg (1506 lb), and carcass weights ranged from 207 to 453 kg (999 lb). All animals were classified as A carcass maturity (9 to 30 months of age).

The task of obtaining the exact number of animals needed for an experimental study of this type is practically impossible. Marbling scores and actual percentage of intramuscular fat (marbling) can only be determined once each animal has been slaughtered, dressed, chilled, USDA-quality graded, and after ether extraction tests were performed on the excised samples of beef carcass LD muscle. The sample distribution of animals targeted for this study was to represent a uniformly distributed profile of 10 to 20 animals

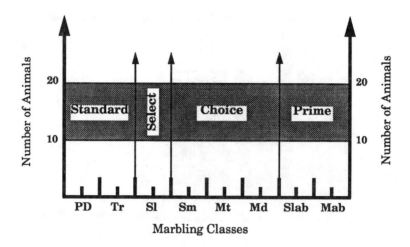

Figure 2.2 Targeted sample distribution with a range of 10 to 20 animals throughout each USDA-quality grade. (From Thane, 1992. With permission.)

representing the USDA-quality grades across each marbling category for A carcass maturity as illustrated in Figure 2.2.

Because of the limited number of animals that fell within the high marbling categories (Slightly Abundant, Moderately Abundant, and Abundant) and low marbling categories (Devoid and Practically Devoid), the target sample distribution was unobtainable. A market profile based on the average numbers of beef animals representing USDA-quality grades across each marbling category for "A" carcass maturity is shown in Figure 2.3. Sample distributions of the 71 live animals and the 88 slaughtered carcasses for this study are shown in Figures 2.4 and 2.5, respectively. The actual number of animals used in this study was further reduced to 62 live animals and 79 slaughtered carcasses. Animals that fell below 2 percent fat content and above 8.5 percent fat content were eliminated from the study. This was done because of the limited number of animals that fell within the ether extractable fat concentration range of 2 percent and below and 8.5 percent and above.

2.1.2   Example: Sampling for detection of peanut off-flavors

Osborn et al. (in press) developed an instrumental method to detect peanut off-flavors. Peanuts from the 1996 season were dug from border rows of experimental plots. The peanuts were shaped into inverted windrows and cured in this orientation for approximately 20 h. Ambient temperature ranged from an overnight low of approximately 16°C to a daytime temperature of approximately 24°C. No rain was observed during this time period. Approximately 10 kg of pods were hand picked from windrows. The moisture content of the pods was approximately 20 percent wet basis when removed from the windrow. The pods were placed in a burlap bag and transported to the lab. The pods were rinsed in tap water to remove soil and patted dry with paper towels.

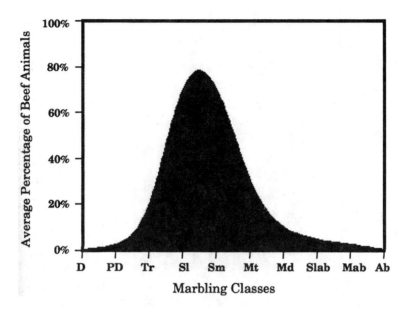

Figure 2.3 Typical market profile of the average number of beef animals represented according to marbling class for each USDA-quality grade. (From Thane, 1992. With permission.)

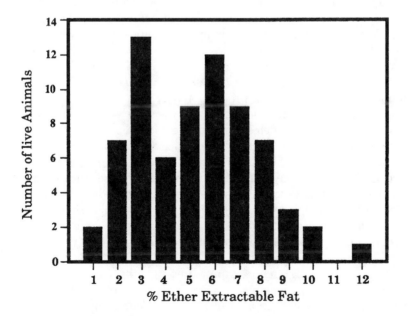

Figure 2.4 Sample distribution of 71 live beef animals. (From Thane, 1992. With permission.)

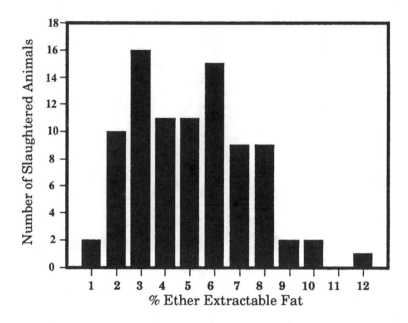

Figure 2.5 Sample distribution of 88 slaughtered beef carcasses. (From Thane, 1992. With permission.)

The maturity of a random sample of the peanuts was determined using the hull scrape method (Williams and Drexler, 1981; Henning, 1983).

The pods were thoroughly mixed and separated into two lots of equal mass. One lot was placed in a gravity convection-drying oven with an air temperature of 50°C. This temperature was higher than reasonably expected in typical peanut curing operations. This temperature was selected to ensure off-flavor volatiles were produced. Osborn et al. (1995) indicated that the potential for off-flavor volatile production decreases with the decrease in peanut moisture content and increases with the increase in temperature. The peanuts were dried to 20 percent moisture content wet basis under ideal field conditions that were not expected to produce any off-flavor volatiles.

The high temperature curing treatment using a temperature slightly above 35°C may produce off-flavors in peanuts at high moisture content but not in peanuts at 20 percent moisture content. The pods were cured in the oven for 65 h. The pods were oven dried from an initial moisture content of approximately 20 percent wet basis to a final moisture content of 6 percent wet basis. The pods were then removed from the oven, cooled to room temperature, and placed in burlap bags for storage at room conditions. This lot is referred to as high temperature cured peanuts.

The remaining lot was placed in a single layer on a countertop in the lab at room temperature (approximately 23°C). These peanuts were allowed to cure for eight days until a hand-held Dickey-John (Auburn, IL) moisture meter indicated a steady moisture content reading. The pods were room dried from an initial moisture content of approximately 20 percent wet basis to a final moisture content of 6 percent wet basis. This lot is referred to as room temperature cured peanuts.

The two lots of peanuts were dried to the same final moisture content to avoid water vapor effects being responsible for different sensor output readings. Both lots of cured peanuts were stored in nonairtight burlap bags in separate cabinets at room conditions for 10 weeks to allow for moisture equilibration. The condition of the air in the room during storage was such that no further drying of the peanuts occurred. The moisture content of the pods, kernels, and hulls was determined after storage using an oven method specified by ASAE standard S410.1 (ASAE, 1998): 130°C for 6 h in a forced air convection oven.

After moisture equilibration and determination, the peanuts were analyzed using the GC (gas chromatograph) and OVM (organic volatiles meter) to verify high concentrations of off-flavor volatiles in the high temperature cured peanuts and low concentrations in the room temperature cured peanuts. GC analysis was performed on 3 different 100 g random samples of kernels removed from both the high temperature cured lot and the room temperature cured pods. The kernels were ground in water and sampled according to the method of Singleton and Pattee (1980, 1989). Data were reported in units of micromoles of volatile compound per kg kernel dry matter. The OVM was used to determine the HSVC (Head Space Volatile Concentration) in parts per million according to the method developed by Dickens et al. (1987). The three OVM tests were performed on the room temperature cured peanuts, and four tests were performed on the high temperature cured peanuts. Each test used 100 g of ground kernels.

In the experimental sampling, each curing temperature treatment lot of pods was thoroughly mixed and separated into 10 replicant sublots of 45 g each. The sample sublots were placed into paper bags and labeled. The order of sublot testing was randomized.

The testing began by shelling the pods in the sublot to be analyzed, removing the redskin from the kernels, and then grinding the kernels using a rotary blade-type coffee grinder for 10 sec without water. Once the hulls were removed from the sublot, the sample size was reduced to approximately 5 g of kernels. The ground kernels were then placed into instrument sampling bags. The filled bag was air conditioned to 25°C and 40 percent relative humidity by the controller within the equipment. This air condition was very near moisture equilibrium conditions for the kernel. The sample plastic bag was then sealed and allowed to equilibrate in the equipment holding chamber at the test temperature (25°C) for 15 min. The reference test air temperature was set to 25°C and the relative humidity set to 30 percent.

2.1.3 Example: Sampling for meat quality evaluation

For a project of meat attributes prediction, 30 beef carcasses were available for study (Moore, 1996). The 30 beef carcasses were obtained from beef animals of known genetic background from three breed types ($\frac{3}{4}$ Angus × $\frac{1}{4}$ Brahman, $\frac{1}{4}$ Angus × $\frac{3}{4}$ Brahman crosses, and F_2 Angus × Brahman crosses). From each of the three breed types, five steers and five heifers were slaughtered at a constant age at the Meat Science and Technology Center of Texas A&M University. Each carcass was split, and the right side of each carcass was electrically stimulated (ES). Electrical stimulation was achieved using a Lectro-Tender™ electrical stimulation unit (LeFiell Company, San Francisco, CA).

The ES procedure was performed within 1 h postmortem and consisted of 17 impulses of 550 volts and 2 to 6 amps for 1 min. Impulses were 1.8 sec in duration with 1.8 sec of rest between impulses. At 24 h postmortem, approximately 50 g of muscle tissue were removed from 4 muscles on each side of the carcasses [semimembranosus (Sm), semitendinosus (St), triceps brachii (Tb) and biceps femoris (Bf), and the longissimus muscle from the nonelectrically stimulated side for calpastatin enzyme activity and sarcomere length measurements]. At 48 h postmortem, 2 trained personnel collected USDA-quality and yield grade characteristics (hot carcass weight; percent kidney, pelvic, and heart fat (KPH); ribeye area (REA); fat thickness opposite the 12th rib; adjusted fat thickness; marbling score; lean maturity; skeletal maturity; quality grade and yield grade) (USDA, 1989).

Each carcass then was fabricated, and the nine major muscles were removed. A total of four steaks 2.54 cm thick were cut from the anterior end of each muscle for Warner–Bratzler shear force determination and assigned to 1 of 4 aging periods: 2, 14, 28, and 42 days postmortem. Steaks were stored at $4 \pm 2°C$ until completion of the respective aging period. An approximately $80 \times 80 \times 50$ mm block was taken from the posterior end of each muscle at 2 days postmortem for elastography analysis (samples were not taken from the nonelectrically stimulated (NS) LD owing to the limited size of the muscle allocated for this study). The remaining tissue was ground for percent fat, percent moisture, collagen amount, and solubility analysis. Samples were vacuum packaged in high oxygen barrier bags (B540 bags; Cryovac Division, W.R. Grace & Co., Duncan, SC) and frozen at $-10°C$ until the analyses could be conducted.

2.1.4   Example: Sampling for snack food eating quality evaluation

In order to evaluate the eating quality of snacks and chips, a project was planned to develop an objective method through quantization of the quality characteristics of the products (Sayeed et al., 1995). For this study, a considerably large number of samples (at least 600 bags, roughly 100 samples per bag) of the snacks were prepared. These samples were produced under different process and machine wear conditions. In addition, the raw material was varied to evaluate its effect on the snack quality. Each of the machine wear–raw product scenarios was tested under different (at least 10) process conditions referred to as "cells." The samples for each of the cell conditions were sealed in different bags. For this study, five bags of samples under the same cell condition were mixed in order to obtain a representative sample. Then, samples were collected randomly under the same cell conditions.

2.1.5   Example: Sampling for snack food frying quality process control

The data from the process were generated around typical setpoints by perturbing the process with a pseudo random binary signal (PRBS). The PRBS test, also known as a dither test, was designed for the identification of the input–output models of the process. A PRBS is easy for linear process modeling. A PRBS that contains a random sequence of zeros and ones can be generated by the following codes.

if $r < fc$ then

$$u(k + 1) = u(k)$$

else

$$u(k + 1) = 1 - u(k)$$

end

where r is uniformly distributed on [0 1], fc is a preset threshold value, and $u(0)$ is 0 or 1.

The value of the fc depends on the sampling rate, the dynamics of the process (fast or slow), and the frequency components. It represents the chance the PRBS flips from 0 to 1 or from 1 to 0. The smaller the value of fc, the lower the frequency components in the data from the process are collected. In this study, low frequency components are concerned because process models are needed to be able to do long-term prediction. Therefore, for the fryer, the sampling time was set at 5 sec, and the fc value was set in the range of 0.125 and 0.15. With the PRBS, the inlet oil temperature, the submerger speed, and takeout speed were each dithered between 3 independent setpoints for approximately 6500 time steps as the process input.

The exogenous inputs are setup values, not actual values. For example, in the frying process, the setup values for temperature are provided to a PID (proportional–integral–derivative) controller, which then controls the oil inlet temperature. It is important to know the relationship between the setup input and actual input in order to model the process correctly.

It also is important to have suitable experimental conditions and to keep the conditions stable in order to acquire valid data to build effective models for the process. In this case, the temperature in the plant has such an effect on the frying process that the test should be done when the plant temperature is constant in order to make the process stationary. Further, before the test starts, it is imperative that the frying process is in a steady state, otherwise the process becomes nonstationary. In this way, the stationary data acquired from the test is able to facilitate the analysis, modeling, and controller design of the process.

2.2   Concepts and systems for data acquisition

When samples are selected and prepared for study, it is time for sample data acquisition. Currently, a typical DAQ system is based on a personal computer (PC) in connection with sensors or transducers, signal conditioning, DAQ hardware, and software. In laboratory analysis, industrial measurement, test, and control, personal computers with expansion buses are being used for data acquisition. Figure 2.6 describes the typical structure of such a DAQ system. In the system, depending upon the design of the components, the PC, transducers, signal conditioning box, DAQ hardware box, and software are integrated to produce the desired data.

In such a DAQ system, the PC is the center. It co-ordinates the operation of the system from data sensing, signal conditioning, A/D (analog to digital) conversion, and signal analysis and processing.

The physical properties of an object are measured by sensors. Then, the transducers convert the measurements into electrical signals which are pro-portional to the physical parameters. For example, in the food industry temperature is the most common physical parameter that is measured. In PC-based DAQ systems, thermocouples, resistance temperature detectors (RTD), thermistors, optical pyrometers, or solid-state temperature sensors may be used to convert temperature into a voltage or resistance value. Other common sensors are strain gauges to measure changing length and pressure transducers and flow transducers that transform pressure and flow rate to electrical signals. Other sensors and transducers measure rheological prop-erties, level, pH and other ions, color, composition, and moisture.

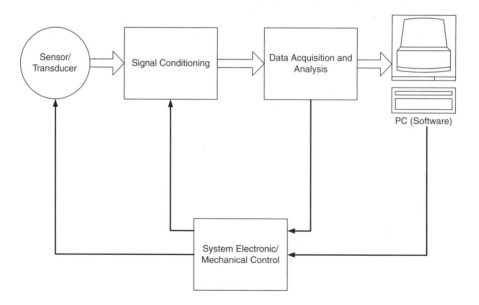

Figure 2.6 A typical DAQ system.

Often the terms "transducer" and "sensor" are used interchangeably. However, there is a distinction that must be made between them. Sensors are devices or elements that interact with a desired measurand with a response that is detectable. For example, a thermistor is a sensor that changes resistance in response to changes in the thermodynamic energy (i.e., temperature) of the surrounding fluid. While the sensor creates a measurable response, all DAQ systems require a voltage input at the point of A/D conversion. Thus, the transducer is comprised of a sensor and circuitry that converts the sensor output to a voltage. The transducer may also be designed to scale (amplify or attenuate) the signal, linearize the output, or change the signal in some other manner. It is also possible that the conversion of the signal occurs partially at the sensor and partly at the DAQ. For example, many industrial grade transducers produce a 4 to 20 mA output signal. Current signals have the advantage of being relatively immune to electrical noise in plant environments and free of resistance changes in the electrical wires used to connect the transducer to the DAQ. Because the DAQ is often enclosed in an instrument room or control cabinet some distance from the transducer, small changes in wire resistance can introduce significant error. In this case, the current signal is transformed to voltage at the DAQ through a relatively simple circuit. There are specialized DAQ systems designed to accept direct sensor inputs and perform the signal conversions at the DAQ.

Signal conversion in the transducer can be thought as part of the signal-conditioning element. Typically, signal conditioning is an analog process and may include signal amplification, attenuation, linearization, impedance matching, and filtering. Signal amplification involves increasing the amplitude of the signal. This generally improves the signal to noise ratio (SNR) and can enhance resolution in DAQ. As will be discussed further, the resolution of the A/D conversion is determined by the number of digital bits and the range of the reference voltage and is a fixed value. For a 12 bit A/D with a 0 to 10 V reference, the resolution is 2.44 mV regardless of the amplitude of the input signal. Thus, if the signal is amplified before entering the A/D converter, the error owing to A/D resolution becomes a smaller fraction of the signal. Signal attenuation is the reduction of the signal amplitude and may be applied when the input is greater than the reference voltage on the A/D converter.

Linearization is another common type of signal conditioning. Often the output of a transducer is nonlinear with respect to the input. A typical example is the thermistor, a device whose resistance decreases exponentially with respect to temperature. While the correct reading can be calculated from a known mathematical function, it is often preferred to convert the nonlinear response to a linear output using analog electronics. Analog linearization is faster than calculating the conversion on a digital signal and is well suited to a direct readout.

Impedance matching is an important type of signal conditioning that is sometimes overlooked. As a general rule, the output impedance of any stage of the transducer should be at least 1/10 of the input impedance of the DAQ.

This minimizes the current flow through the DAQ and subsequent errors in reading the voltage at the DAQ. If, for example, the output impedance of the transducer and the input impedance of the DAQ were equal, then approximately one-half of the current would flow through the DAQ and one-half through the transducer circuit. This would result in a voltage reading at the DAQ that was approximately one-half of the true voltage.

Filtering is not usually part of commercial transducers (i.e., sensors plus signal conditioning) but sometimes added to improve the measurement process. Signal filters are designed to attenuate some frequencies contained within a signal while allowing others to pass through relatively unchanged. A low pass filter allows low frequencies to pass through the circuit to the DAQ while reducing the amplitude of the higher frequency signals. High pass filters block low frequency signals, band pass filters allow only signals within a certain frequency range to pass, and notch filters block signals in a set frequency range. In many applications, electrical interference at a frequency of 60 Hz is pronounced. If the signals we are interested in occur at a lower frequency, for example, a temperature change that occurs less than once per second (1 Hz), then a low pass filter can be designed to block the 60 Hz noise while passing the 1 Hz signal. Analog filter design is covered in most elementary electrical engineering texts and is beyond the scope of this book. The signal of interest and the surrounding electrical interference determine which type of filter is needed for a given application.

Biosensors are used in food analysis and process control. Biosensors are devices joining molecular biology and electronics. They are called biosensors because they use biomaterial as the sensor component and have electronic components to modify and condition the detected signal. Generally, a biosensor consists of a biological receptor and a transducer. The biological component is close to, or fixed on, the surface of the transducer. The transducer converts the biological reactions from the output of the receptors into a response that can be processed as an electrical signal. Biosensor techniques originated in the field of medical science, but more applications are seen in all stages of food production, from raw materials through processing. For example, fruit ripeness, pesticide and herbicide levels of vegetables, and rapid microbial change in dairy products, fish, and meat have all been measured by biosensors. Interested readers can refer to the works of Taylor (1990), Wagner and Guilbault (1994), and Scott (1998).

Understanding the specifications of the DAQ system is important in establishing the capability and the accuracy of the system. DAQ parameters include the sampling rate per channel, voltage resolution, the input voltage range, and the number of channels for analog input. In a PC-based DAQ system, the rate at which samples are taken is important. The sampling rate can be expressed either in the sampling interval, that is, the time between samples or more typically by the sampling frequency (samples per second). Sampling frequency can be misleading because a manufacturer may quote a sampling frequency across all analog input channels. For example, a DAQ that may be specified as sampling at 8 kHz and having 16 analog input channels may only sample each channel at a rate of 500 Hz.

The required sampling rate is determined by the frequency content of the signal that is to be recorded. The Nyquist criteria specify that the sampling frequency must be at least twice the highest signal frequency. Otherwise, the signal will be distorted and appear to be at a completely different frequency. This phenomenon is known as aliasing. A good rule of thumb in selecting the sampling rate is to use a rate that is approximately 10 times the desired frequency, but note that signals with a broad frequency range may require a low pass filter prior to the D/A step to prevent aliasing which would create error in the values at lower frequencies.

The voltage resolution is determined by the method of A/D conversion. There are three basic methods used in A/D converters: successive approximation, flash, and dual slope. Each works on a different basis, but successive approximation A/D is far and away the most commonly used. In a successive approximation A/D converter, the number of bits used to represent an analog voltage and the reference voltage of the converter determine the resolution by the formula

$$\Delta D = \frac{V_{ref}}{2^n}$$

where ΔD is the resolution in the digital number, V_{ref} is the reference voltage, and n is the number of bits in the converter. Thus, a 12-bit converter with a reference voltage of 10 V will have a resolution of 2.44 mV.

Input voltage range is established by the design of the A/D converter and can be unipolar (e.g., 0 to 10 V) or bipolar (e.g., -10 V to 10 V). Less expensive DAQ equipment may only have a single input range, but more expensive equipment will provide a selection of ranges. It is generally good practice to select a range that best matches the input voltage from the transducer.

Only one analog input voltage at a time can be processed per A/D converter. However, many applications require the DAQ system to process multiple voltage inputs. This is usually accomplished by using a multiplexer (MUX). The MUX switches between incoming signals and feeds each in turn to the A/D for processing. Modern DAQ systems usually are built to accommodate 8 to 16 channels, and with additional hardware that number may be increased to 96 or more. However, increasing the number of channels beyond those contained in the A/D board decreases the sampling frequency per sample.

Image signals require a special interface card, usually referred to as an image acquisition card, or a "frame grabber," in combination with a camera or other image-producing device in order to capture digital images.

Measurement and data acquisition techniques are in development to measure very high viscosities, moisture content, and color of uneven surfaces. Nondestructive, real-time measurement of enzyme activity, microorganism activity, flavor/aroma, and shelf life will enhance and ensure food quality, consistency, integrity, and safety. The following examples of system setup are ultrasonic A-mode signal acquisition for beef grading, an electronic nose which emulates human nose sensing peanut flavor, and on-line measurement for color, moisture content, and oil content of snack food products.

2.2.1   Example: Ultrasonic A-mode signal acquisition for beef grading

Meat palatability is often related to the percent of intramuscular fat present in a meat cut. In fact, the USDA-quality grades are based primarily upon "marbling," the visual appraisal of intramuscular fat on a cut surface of a chilled carcass between the 12th and 13th ribs (USDA, 1997). Other factors that influence palatability include animal age, amount of connective tissue, and aging effects. However, intramuscular fat is the primary component. Park and Whittaker (Park and Whittaker, 1990 and 1991; Park, 1992; Whittaker et al., 1992) conducted an overall study focused on developing a robust method for measuring the amount of intramuscular fat in a beef cut non-invasively. This section describes their work in setting up the ultrasonic A-mode signal acquisition system for beef grading.

There are several types of ultrasonic diagnostic instruments, such as A-mode, B-mode, C-mode, M-mode, Doppler devices, etc. (Christensen, 1988). Among them the A-mode technique is the oldest. A-mode ultrasound gives one-dimensional information, which shows the amplitude of the signal. Beef quality is generally evaluated using a marbling score as an indicator of the percentage of intramuscular fat in the LD muscle. Because the ultrasonic speeds are different between fat and muscle, the marbling scores can be predicted by measuring the ultrasonic longitudinal and shear speed in biological tissue.

Longitudinal and shear ultrasonic transducers (1 MHz, 2.25 MHz, and 5 MHz) were used to obtain data from each beef sample. The longitudinal acoustic velocity and attenuation were recorded from the signal in time domain. Contact transducers (Panametrics, Waltham, MA), 12.7 mm in diameter, with longitudinal wave frequencies of 1 MHz, 2.25 MHz, and 5 MHz were used. An ultrasonic analyzer (Model 5052UA, Panametrics) that incorporated a broadband pulser–receiver and a gated peak detector were used to measure ultrasonic speed and attenuation.

The signal from the ultrasonic analyzer was then acquired through an 8-bit, 20-MHz A/D converter board (PCTR-160, General Research, McLean, VA) and processed by software (PC-DAS, General Research) that enabled ultrasound speed and attenuation to be measured and the results displayed on the monitor screen. The peak amplitude measurement method was used to quantify attenuation. Figure 2.7 shows a schematic diagram for the experimental set-up, and Figure 2.8 shows the diagram of system blocks.

Theoretically, the sensitivity of the speed to fat percentage of the longissimus muscle was 1.6 m per (s minus the percent of fat). In fact, the average distance the ultrasound wave traveled through the meat sample was 25.9 mm

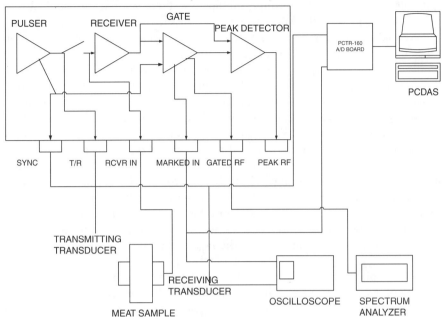

Figure 2.7 Schematic diagram for the experimental setup for measuring ultrasonic speed and attenuation. (Adapted from Whittaker et al., 1992. With permission.)

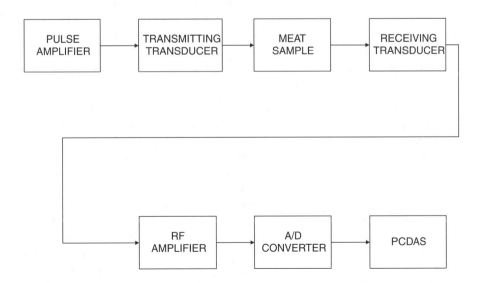

Figure 2.8 Block diagram for experimental setup for measuring ultrasonic speed and attenuation. (Adapted from Whittaker et al., 1992. With permission.)

for excised meat samples that were nominally 30 mm. These samples were put into a sample holder (aluminum block) to be measured ultrasonically. In this case, the sample thickness was rescaled using a micrometer for accurate ultrasonic speed measurement. Assuming an average velocity of propagation of 1,500 m per s, the time of flight of an ultrasonic pulse in a section 2.59 cm thick was about 17.27 μs. The average minimum difference of fat between traces and practically devoid marbling scores was approximately 0.71 percent. Assuming a decrease in velocity of 1.14 m per s because of the increase in fat content going from practically devoid to traces, the time of flight in a 2.59 cm thick meat section would be 17.28 μs. Therefore, the resolution required for measuring the arrival time of the ultrasonic signal was 0.01 μs.

Ultrasonic speed measurements were taken at 22 to 24°C temperature in thawed beef samples that were previously frozen. The measurements were obtained using the normal incidence pulse–echo method. In this configuration, the incident beam strikes the interface, resulting in a reflected beam returning in the opposite direction in the same material and a refracted beam passing through the interface into the next material. Ultrasonic energy that is reflected and returned to the probe is the source of the indications shown on the monitor. All ultrasonic energy that travels completely through the test piece was reflected at the end by a back reflector, giving a large back-echo indication. This technique could be used in scanning hot carcasses and live beef animals, provided that either

1. The ultrasonic speed of the medium is known.
2. The exact distance to the back reflector can be determined.

The ultrasonic signal obtained in the DAQ system was further preprocessed, analyzed, modeled, and classified to establish a robust method for measuring the amount of intramuscular fat in a beef cut noninvasively.

2.2.2   Example: Electronic nose data acquisition for food odor measurement

The process of olfaction in mammals is useful to consider in the development and application of an electronic nose. The fundamental problems in olfaction are establishing the limits of detection, recognizing a particular odorant, coding the response to establish odor intensity and quality, and discriminating between similar odorants. From these problems, the need to understand the nature of the olfactory chemoreceptors, the mechanisms of transduction, the neural coding mechanisms for intensity and quality, and the nature of the higher information processing pathways follows (Persaud et al., 1988). Figure 2.9 shows a schematic of the olfactory anatomy in humans.

Perception of odor in humans is not well understood. Odorant molecules must reach the olfactory epithelium at the top of the nasal passages where the olfactory receptors are located. Transport to the epithelium is turbulent and results in a nonlinear relationship between concentrations, flow rate, and the number of molecules reaching the membrane. Approximately 5 to

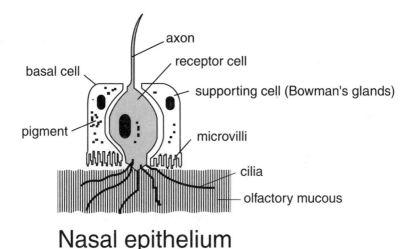

Nasal epithelium

Figure 2.9 Schematic of the human olfactory organs. (From Lacey and Osborn, 1998. With permission.)

20 percent of the air inhaled will pass through the olfactory cleft, depending on the nasal flow rate. Higher nasal flow rates correspond to increased perceived odor intensity. In addition to transport via the nasal flow, humans possess a retronasal pathway from the mouth to the olfactory epithelium that allows odorant molecules to reach the olfactory receptors by a second pathway. This accounts for much of the sensory perception attributed to flavor when eating. Once odorant molecules reach the epithelium, they must be dissolved into the mucus that covers the olfactory epithelium. The dissolved molecules are transported to the receptor cells and their cilia where they interact with the receptors and are converted to neural signals. Volatile molecules are also capable of stimulating the trigeminal nerve endings. While this effect is not fully understood, it is believed that the trigeminal nerves also play an important role in odor perception. Transduction and coding of the receptor response to odorants is not easily reduced and there is no clearly defined theory that covers the perception of the odors (Engen, 1982; Lawless, 1997). The olfactory system responds nonspecifically to a wide variety of volatile molecules. However, most people are able to easily recognize a variety of odors even at very low concentrations.

The electronic nose in Figure 2.10 can be viewed as a greatly simplified copy of the olfactory anatomy, seen in Table 2.1. The receptor cells and cilia are replaced with nonspecific gas sensors that react to various volatile compounds. Because there is generally no mucus into which the odorants must dissolve, the molecules must adsorb onto the sensor. There are a variety of sensors that have been employed including those based on metal oxides (Brynzari et al., 1997; Ishida et al., 1996; Egashira and Shimizu, 1993; Nanto et al., 1993), semiconducting polymers (Buehler, 1997; Freund and Lewis, 1997;

Table 2.1 Comparison of Human Nose vs. Electronic Nose

Item	Human Nose	Electronic Nose
Number of olfactory receptor cells/sensors	40 million	4 to 32
Area of olfactory mucosa/sensors	5 cm^2	1 cm^2
Diameter of olfactory receptor cell/sensor	40–50 micron	800 micron
Number of cilia per olfactory receptor cell	10–30	0
Length of cilia on olfactory receptor cell	100–150 micron	N/A
Concentration for detection threshold of musk	0.00004 mg/liter air	Unknown

Adapted From Lacey and Osborn, 1998. With permission.

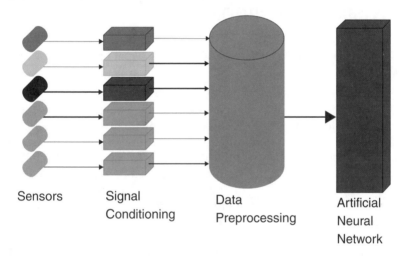

Sensors   Signal   Data   Artificial
               Conditioning   Preprocessing   Neural
                                                   Network

Figure 2.10 Schematic of generic electronic nose. (From Lacey and Osborn, 1998. With permission.)

Yim et al., 1993; Miasik et al., 1986), optical methods (White et al., 1996; Collins and Rose-Pehrsson, 1995), and quartz resonance (di Natale et al., 1996; Matsuno et al., 1995; Moriizumi et al., 1994). Metal oxide and semiconducting polymer sensors are the two most commonly used sensors in commercial instruments. Generally, steady state sensor response has been used in electronic nose systems (Egashira, 1997; Brailsford et al., 1996), but research indicates that transient response measurements can enhance the ability of the sensors to differentiate between different molecular species (Eklov et al., 1997; Llobert et al., 1997; Nakata et al., 1996; Amrani et al., 1995).

Transduction of the olfactometry receptors is replaced with signal conditioning circuits that involve a conversion to voltage (Corcoran, 1994). Coding of the neural signals for odor intensity and odor recognition in humans is replaced with some type of pattern recognition method (Wide et al., 1997;

Hanaki et al., 1996; Moy and Collins, 1996; Ping and Jun, 1996; Davide et al., 1995; Yea et al., 1994; Nayak et al., 1993; Sundgren et al., 1991).

Despite the limitations of the electronic nose as compared to human olfaction, there have been a number of reports of successful application of the electronic nose to problems in foods (Gardner et al., 1998; Lacey and Osborn, 1998; Osborn et al., 1995; Bartlett et al., 1997; Simon et al., 1996; Strassburger, 1996; Vernat-Rossi et al., 1996; Pisanelli et al., 1994; Winquist et al., 1993).

There are sources of error in an electronic nose. Many of them are the same as the error sources in olfactometry. These errors include a lack of sensitivity to odors of interest, interference from nonodorous molecules, effects of temperature and humidity, nonlinearity of the sensor response, and errors from sampling methodology. It is beyond the scope of this book to discuss these errors as they affect olfactometry measurements. However, there are several references to quality control in olfactometry (Berglund et al., 1988; Williams, 1986) and published standards for olfactometry measurements (ASTM, 1991; ASTM, 1988).

Osborn et al. (in press) developed an application of a commercial electronic nose for detecting high temperature curing off-flavors in peanuts. Peanuts were tested in four progressive states of destruction: whole pods, whole kernels with red skins, half kernels without red skins, and ground kernels. Off-flavors in ground kernels were also measured using GC and an OVM for comparison to the electronic nose. The electronic nose sensor array was able to separate good from off-flavored peanuts after some data processing to remove bias effects from an unknown source. The bias was suspected to come from slight water vapor differences between the samples that affected all sensors equally. Further, the electronic nose was able to differentiate between the samples nondestructively suggesting that there may be a potential to use this technique to establish quality control points in the processing that could reduce or eliminate off-flavor peanuts.

2.2.3   Example: Snack food frying data acquisition for quality process control

The data used for the study of snack food frying quality process control were acquired from a frying process described in Chapter 1. The food materials were dropped into a fryer by a conveyor. Then, a submerger conveyor kept the materials immersed in the oil and pushed them forward through the fryer. The submerger conveyor speed was changed to adjust the residence time of the product in the fryer.

A takeout conveyor moved the products out of the oil through a series of paddles. The products were then transported by a conveyor to a cooling section which was open to ambient conditions. After this, the product passed over a weigh belt where the sensors for product quality attribute measurements were located. The quality attributes measured on-line were color, moisture content, and oil content of the final products.

The color was measured with a Colorex sensor (InfraRed Engineering, Inc., Waltham, MA) placed after the cooling section. The sensor measured color CIELAB units in $L*a*b$ space. The CIELAB scale is preferred (over the CIELUV scale) by those who work with products that have been pigmented or dyed (HunterLab, 1995). This scale correlates more closely with the manner in which humans actually see color because it is based on the opponent colors theory of vision (HunterLab, 1995). This theory assumes that humans do not see red and green or yellow and blue at the same time. Therefore, single values can be used to express the red or green and yellow or blue attributes of objects.

The b value of the CIELAB scale is positive when the product is yellow and negative when the product is blue. In other words, b is a measure of yellowness–blueness in the product. In this work, the b value was used over a and L for modeling and control because it was affected the most when the process was varied.

Moisture content is the quality of water per unit mass of either wet or dry product, wet basis, or dry basis, respectively. In this case, it was done in a wet basis. The oil content measurements were also performed in a wet basis. Moisture content and oil content were measured with a MM55 sensor (InfraRed Engineering, Inc., Waltham, MA) located after the cooling section.

For the purposes of modeling and control, the frying process was simplified to an input–output structure (Figure 2.11) by studying the process mechanism. In this structure, the inlet frying oil temperature, the speed of the submerger conveyors, and the speed of the takeout conveyors were chosen as the process input variables (independent variables in the context of modeling or manipulative variables in the context of automatic control). These input variables had a strong impact on the process output variables (also known as the dependent variables or controlled variables) that are the product quality attributes: color b, moisture content, and oil content.

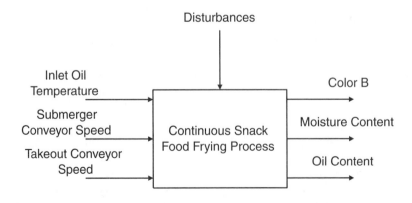

Figure 2.11 A simplified input–output structure of the snack food frying process.

This structure is the basis of data acquisition, analysis, modeling, and control of the process.

2.3 Image acquisition

In recent years, techniques of image acquisition, analysis, and pattern recognition have merged with computer technology and increasingly have been applied to problems in food quality evaluation and process control. Current imaging methods rely on digital images to visualize physical properties of food or ingredients in order to quantify parameters for analysis and process control. Applications in food engineering mostly have involved the evaluation of food quality based on the information derived from image texture, product morphology (size and shape), and product color features extracted from digital images of food samples. Images have been developed from various sensors capable of visualizing the food samples such as digital cameras measuring visible light reflectance; detection devices for ultraviolet (UV), infrared (IR), and near infrared (NIR) reflected radiation; X-ray transmission; ultrasound transmission; and nuclear magnetic resonance (NMR). A computer is usually needed to process and analyze the acquired images. In an imaging system, the computer is analogous to the human brain and the sensor is analogous to human eyes. An imaging system can be viewed as a simulation and extension of the human vision system and, in many cases, imaging systems are also called machine vision systems.

The signals from the imaging sensor are analog. The analog imaging signals are sent to an A/D converter in order to transform them into digital signals. Then, the digital signals are formed directly into images or are manipulated with a mathematical transformation to a final image. After that, images are stored. The process of image transforms and formation is generally referred to as image processing and is an important aspect for application development. In general, image processing for machine vision systems includes the following steps:

1. Load the raw images from the storage.
2. Enhance and segment the images for desired information.
3. Extract textural, morphological, and/or color features of the images.

The extracted features can be used to classify the samples by quality parameters. The results can be fed back to adjust the image acquisition for better performance of the system. This section focuses on image acquisition while image processing and classification are discussed in the following chapters.

The structure of a generic machine vision system is shown in Figure 2.12. In machine vision systems, the most common imaging sensor is a digital camera. The digital camera, in connection to a computer, takes pictures directly from the food samples with the help of illumination from different light sources. This imaging sensing method performs visual detection of external features of the food samples. Recently, there have been more exotic imaging

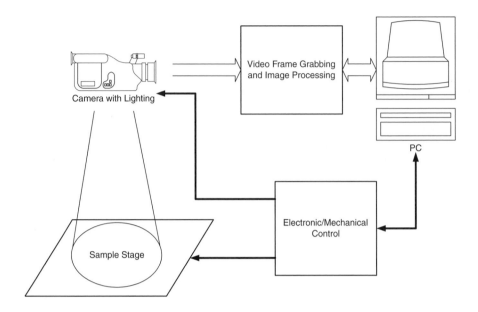

Figure 2.12 Structure of generic machine vision system.

methods applied to food quality evaluation and process control. Many of these imaging methods originate from the medical profession. Dual energy X-ray (Ergun et al., 1994), ultrasonic B-mode (Christensen, 1988) and elastography (Ophir et al., 1991), NMR and MRI (Magnetic Resonance Imaging) (Ruan et al., 1991; Zeng et al., 1996), are a few examples.

These imaging methods are especially effective in detection of internal characteristics within food products. An example is ultrasonic elastography. This technique was invented for diagnosis of breast cancer (Ophir et al., 1991) where tumors within soft tissue were difficult to detect with X-ray images but differences in tissue elasticity yielded better detection with ultrasound combined with tissue compression. Elastography was later applied to the prediction of meat quality attributes (Lozano, 1995; Moore, 1996; Huang et al., 1997). Imaging methods utilizing different sensors generally differ only in image acquisition and formation. The algorithms for imaging processing and classification are basically the same.

2.3.1   Example: Image acquisition for snack food quality evaluation

Sayeed et al. (1995) developed a machine vision system for quantization of snack quality in order to develop a methodology that was useful in the snack food industry to evaluate product quality in a nondestructive manner. A color camera captured images of a snack product. After image processing, the image texture, size, and shape features were used as input to an artificial neural network (ANN) model to predict the sensory attributes of the snack food quality.

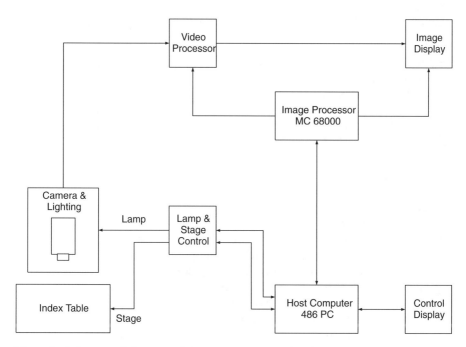

Figure 2.13 Structural diagram of the imaging system for the snack food quality quantization. (Adapted from Sayeed et al., 1995. With permission.)

Figure 2.13 shows a schematic of the image acquisition and processing system. The samples were imaged using a Quantimet imaging system (Quantimet 570 Image Processing and Analysis System, Leica Cambridge Ltd., 1991). The image acquisition system was equipped with a CCD (charged coupled device) color camera that captured multiple frames (as many as 32 images per acquisition) of size 512 × 512 pixels. The imaging system had a lighting setup that could be controlled via software for proper illumination. Before capturing images, all camera and stage parameters were set so that all conditions remained the same in subsequent image acquisitions. An image was captured as follows: 32 frames were digitized per sample, and these samples were averaged to ensure image quality by reducing random noise. In this experiment, 50 sample images were acquired for each cell. For example, one wear/raw material scenario had 16 cells. So, under such a condition, a total of 800 (16 × 50) images were captured. The resolution calibration factor for all images was 0.183 mm per pixel. In this experiment, the cross-sectional images of a typical puffed extruded corn collet were used.

Such acquired images were used for the extraction of textural and morphological features to classify snack sensory attributes. You will see the discussions of the topic in the following chapters.

2.3.2   Example: Ultrasonic B-mode imaging for beef grading

Real-time diagnostic ultrasound is an important technology used by meat science practitioners for quality prediction. The most successful interpretation was performed by human interpretation of an ultrasonic image (Brethour, 1990). A study performed by Whittaker et al. (1992) was intended to develop automated image enhancement and interpretation techniques to remove human judgment from the process of predicting the percentage of fat in ultrasonic images of beef tissues.

If an image is needed for visualization in ultrasonic signal analysis, B-mode, where B represents "brightness," is often used. In B-mode, the amplitude of the signal is represented by a gray-level intensity in an image. The image is a two-dimensional reconstruction of reflecting interface locations. There are several methods for constructing a B-mode image, but in all cases the image represents a plane away from the transducer, as opposed to a ray in the case of A-mode. Real-time imaging also is used often when an image is needed for visualization. It is similar to B-mode with the exception that it is updated at the video frame rate. This allows the operator to observe movement within the target. This B-mode work was designed to enable the technology to be used to predict intramuscular fat.

The image acquisition–transfer system for the experiment consisted of

1. An ultrasonic diagnostic unit (Aloka Model 633, Corometrics Medical Systems, Wallingford, CT).
2. A 3.5 MHz linear array transducer (Aloka Model UST-5024N, Corometrics Medical Systems).
3. A personal computer (IBM 286 Compatible CompuAdd, CompuAdd, Austin, TX) for image capture.
4. A color video monitor (Sony Model PVM-1341 Trinitron Color Monitor, Sony, Park Ridge, NJ).
5. An image processor (Targa-16 Image Processor, Truevision, Indianapolis, IN).
6. A personal computer-network file server (PC-NFS) software package (PC-NFS, Sun Microsystems, Billerica, MA).
7. A communication board (3C503 Etherlink II, 3Com, Santa Clara, CA).

The image processing system consisted of a color workstation (Sun 3/60, Sun Microsystems) and a software package (HIPSPLUS, SharpImage Software, NY) for image processing. The HIPSPLUS software package included HIPS, HIPSADDON, and SUNANIM modules.

The experimental process for this study was categorized into the following phases.

1. Signal prescanning.
2. Signal scanning.
3. Image acquisition.

4. Image preprocessing.
5. Image processing.
6. Beef quality grading.
7. Data analysis.

Live cattle were scanned in squeeze chutes and the slaughtered animals were scanned on the abattoir floor while hanging from the rail by both hind legs. In the scanning phase, a 3.5 MHz linear array transducer was applied to the scanning area between the 12th and 13th rib cross-section. Signal gain was kept constant for all animals. In the acquisition phase, five cross-sectional images of each test animal's longissimus muscle were captured at the same location. These images were saved on data cartridge tapes for later image processing and data analysis.

2.3.3   Example: Elastographic imaging for meat quality evaluation

Elastography is a method for quantitative imaging of strain and elastic modulus distribution in soft tissues. It is based on time shift estimation between ultrasonic A-line pairs, which are signal lines for the ultrasonic A-mode, at different compressions (Ophir et al., 1991). This approach was proposed for use in evaluating meat muscles (Ophir et al., 1994; Lozano, 1995; Moore, 1996; Huang et al., 1997). Images generated by elastography, called elastograms, were correlated with sensory, chemical, and mechanical attributes of the same meat samples to analyze the relationship between meat tissue profiles and meat quality attributes (Lozano, 1995; Moore, 1996; Huang et al., 1997, 1998). This section presents data acquisition and image formation of elastography. The following chapters discuss image processing, modeling, and prediction of meat attributes based on elastographic imaging.

Elastography was originally developed in the medical field for cancer diagnosis (Ophir et al., 1991; Cespedes et al., 1993). It is a technique for making quantitative cross-sectional images of tissue elasticity. Elastography uses ultrasonic pulses to track the internal displacements of small tissue elements in response to an externally applied stress. These displacements are converted into local strain values along the axis of compression by comparing pre- and postcompression signals from within the tissue. The strain values may be converted into calibrated Young's modulus values with the additional knowledge of the applied stress and its distribution along the compression axis. The resultant image is called an elastogram. For practical reasons, elastograms actually display the inverse Young's modulus values, such that lighter regions in the image correspond to the softer, less elastic structures. Figure 2.14 shows an elastogram of beef sample in the LD muscle, in which the compression direction was from top to bottom.

Elastograms have several interesting properties over sonograms (Ophir et al., 1994) which make elastography a suitable method for the evaluation of bovine muscle structure and composition. Elastography combines ultrasonic

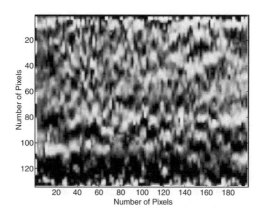

Figure 2.14 An elastogram of beef sample in LD muscle. (From Huang et al., 1997. With permission.)

technology with compression tests resulting in images (elastograms) that carry structural and textural information.

In order to illustrate the principle, consider a simple one-dimensional cascaded spring system where the spring constants represent the elastic moduli of tissue regions and each spring represents the behavior of a cylindrical tissue element with a unit cross section. The total strain will be distributed such that the stress is equal within each spring. For example, if three springs are assumed to have the same modulus, then an axial downward force compresses each spring in the same deformation, one-third of the total deformation. In another situation, if the two outer springs are assumed to have the same modulus and the middle spring a larger modulus, then a force will deform that the outer springs will be the same, but the middle spring will deform less. The strain profile is dependent on the initial compression and on the number and stiffness of all springs. A given local measured value of the strain is influenced by the elastic properties of elements located elsewhere along the axis of compression. For these reasons, it appears that while strain profiling could be useful for imaging, it may be of limited use for quantitative estimation of local tissue elasticity.

It might become possible to estimate the elastic modulus of each component in this system of springs through applying a known stress instead of imparting a known displacement because the stress remains constant with depth in this one-dimensional system. Thus, the measurable strain in each spring and the known stress on each spring could be used to construct an elastic modulus profile along the compression axis. Such a profile would be independent of the initial compression and the interdependence among the component springs would disappear.

In the more realistic three-dimensional cases, however, the applied stress would not be constant along the axis of the compressor. The reason for this lies in the fact that stresses along the transverse direction become important and because their vertical force components are a function of the displacement which, in turn, is a function of the depth and the resultant forces along the compression axis vary with depth. It seems intuitive that by enlarging the area of compression, the transverse springs that are actually stretched and contribute to the depth dependent stress field would become less important. Thus, the applied stress field would be more uniform. Indeed, the theoretical solution to the three-dimensional problem demonstrates this effect. Experiments have also confirmed that a larger area of compression results in a more uniform axial stress field (Ponnekanti et al., 1992).

When tissue is compressed by a quasi-static stress, all points in the tissue experience a resulting level of three-dimensional strain. Although tissue exhibits viscoelastic properties, only the elastic properties are observed. Generally, a rapid compression is applied and the slow viscous properties are ignored. In elastography, a static stress is applied from one direction and the estimation of strain along the ultrasound beam (longitudinal strain) in small tissue elements is measured. If one or more of the tissue elements has a different hardness than the others, the level of strain in that element will be higher or lower. Thus a harder tissue element will experience less strain than a softer one.

The longitudinal component of the strain is estimated from the displacement determined by a time shift measurement assuming a constant speed of sound. This is accomplished by

1. Acquiring a set of digitized ultrasonic A-lines from a region of interest in the tissue.
2. Compressing the tissue (usually using the ultrasonic transducer) along the ultrasonic radiation axis.
3. Acquiring a second postcompression set of A-lines from the same region of interest in the tissue.
4. Performing cross-correlation estimates of time shifts.

Congruent A-lines are windowed into temporal segments, and the corresponding time shifts of the segments are measured using correlation analysis. Accurate estimates of the time shift were obtained using the correlation coefficient function in combination with quadratic peak interpolation. The change in arrival time of the echoes in the segment before and after compression thus can be estimated. The local longitudinal strain at the depth given by the product $i \times (\Delta T \times c \times 2)$, is calculated as

$$s(i) = \frac{\Delta t(i) - \Delta t(i-1)}{\Delta T} \qquad (2.1)$$

where c is the speed of sound in the elastic medium, $\Delta t(i)$ is the time shift between segments in the indexed segment pair, and ΔT is the space between segments. The window is translated along the temporal axis of the A-line, and the calculation is repeated for all depths.

In general, the precision of the time shift estimate improves with increasing segment size. However, in order to improve the axial resolution of the estimate a small segment size is preferred. Because the relative compression and cross-correlation estimate deteriorates with increasing segment size, this degrades the precision of the estimate. Thus, there are two competing mechanisms that affect the precision of the time shift estimate as a function of the segment size.

As noted previously, strain is a relative measure of elasticity because it depends on the magnitude of the applied compression as well as on the elastic composition of the material. Ultimately, it may be useful to obtain an absolute measure of the local elasticity in the tissue. For three-dimensional tissue elements, the parameters of interest are the elastic modulus in three directions and shear modulus. The elastic modulus E_{ij} is defined as

$$E_{ij} = \frac{\sigma_{ij}}{s_{ij}} \tag{2.2}$$

where s_{ij} and σ_{ij} are the strains and stresses in three directions, respectively. In a uniform isotropic medium under uniaxial stress, the elastic modulus or Young's modulus is simplified as

$$E = \frac{\sigma}{s} \tag{2.3}$$

where σ and s are the strain and stress along the compression direction, respectively. Therefore, one can think of elastograms as basically strain images. The strain estimates can be normalized by the estimated local stress to obtain calibrated elastograms. When the tissue is assumed to be uniform and isotropic, and uniaxial longitudinal stress is applied, the calibrated elastograms are images of the inverse elastic modulus.

The internal stress is defined by the boundary conditions and by the structure of the tissue. Therefore, *a priori* estimation of the exact stress field is impossible because it is target dependent. However, the theoretical estimates of the stress distribution owing to the external boundary conditions can be obtained (Ponnekanti et al., 1992). In practice, when the compression is applied with a compressor that is large compared to the dimension of the tissue volume under examination, the stress in the tissue is approximately uniform and uniaxial with a slight magnitude decay with depth. Previous experiments showed that if the width and the depth of the region of interest are close to the dimensions of the compressor, the stress variations owing to external boundary conditions are minimal (Cespedes and Ophir, 1991). Figure 2.15 shows the functional diagram of the formation of elastograms.

Data acquisition of elastography is performed to capture the arrays of ultrasonic A-lines. Each of the arrays of A-lines represents a quantitative cross-sectional image of tissue elasticity. A standard biomedical ultrasonic

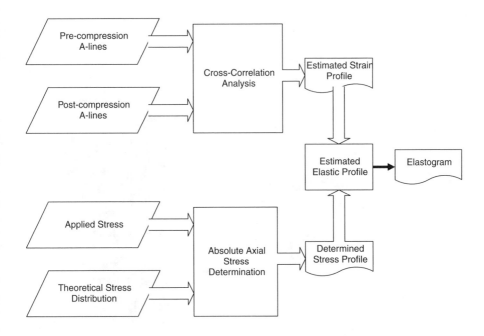

Figure 2.15 Functional diagram of the formation of elastograms.

scanner with a 5 MHz ultrasonic probe is typically used in elastography. The probe is a linear array of ultrasonic transducers where the total number of transducers is 236. Each transducer contributes one A-line and all of the transducers together provide two-dimensional information.

Cespedes et al. (1993) reported a clinical setup of the elastography system. The elastography system consisted of a Diasonic Spectra II ultrasound scanner (Diasonics Inc., Milpitas, CA) equipped with a 5 MHz linear array, a Lecroy 8 bit digitizer (Lecroy Corp., Spring Valley, NY) operating at 50 MHz, a motion control system, a compression device, and a personal computer that controlled the operation of the system. The compression device consisted of a CGR-600 X-ray mammography paddle (GE/CGR, Milwaukee, WI) that was modified to accommodate a transducer holder and a positioning device. Chen (1996) reported a laboratory setup of the elastography system for meat inspection. This system was used to scan a cattle carcass and obtain elastograms of the bovine muscle tissue. The system setup consisted of a mechanical system and an electronic system. The mechanical system included a pneumatic system and a motor and was used to hold the carcass in position and move the transducer probe to make sufficient contact between the probe and the carcass. Figure 2.16 gives a general description of the mechanical system. The pneumatic system provided an up–down motion (location A in Figure 2.16) and an in–out motion (location B in Figure 2.16). The motor carrying the ultrasonic probe was on B and controlled by the motor controller (C). An arm (D) was moved simultaneously with the probe

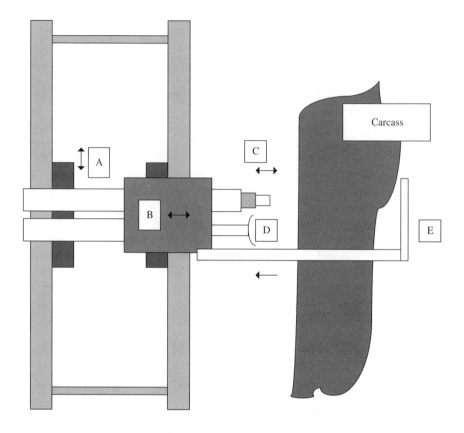

Note: 1. The arrows show the direction of the motion.

2. The moving parts are named   --A, for up/down carriage (pneumatic)
                                  --B, for left/right carriage (pneumatic)
                                  --C, for the probe carriage (motor)
                                  --D, for the carcass front support (static)
                                  --E, for the carcass rear
                                    holder (pneumaitc)

Figure 2.16 Mechanical part of elastographic meat inspection system. (From Chen, 1996. With permission.)

(both are on carriage B) along with the holder (E). The holder had an articulation that was swung on the rear side of the carcass that was now between the arm D and the articulation E. Moving the articulation back immobilized the carcass between D and E. The probe could be moved by the motor over a range of ±75 mm. It could also rotate on its axis so that it could be moved between adjacent ribs to make contact with the meat. The electronic system included a Hitachi ultrasonic scanner, an interface board, a timing control board, a digitizer, a motor control, and a personal computer.

Figure 2.17 gives a general description of the electronic system. A-line radio frequency (RF) signals and the synchronizing control signals were

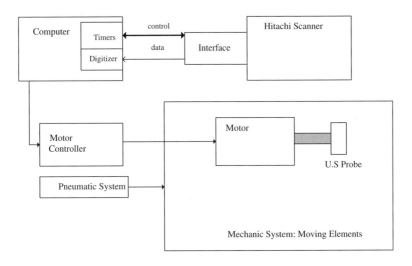

Figure 2.17. Electronic part of elastographic meat inspection system. (From Chen, 1996. With permission.)

obtained from the Hitachi ultrasonic scanner. The signals could be shown by the scanner as a B-scan. The signals were then fed to the interface board. The timing board controlled the data acquisition and the digitizer performed the A/D conversion and temporarily stored the data that was processed by the computer and stored as a file in the hard disk. The digitizer had a sampling frequency of 50 MHz and 8 Mb memory. The signal frequency was around 5 MHz, so the sampling frequency was well above the Nyquist rate. The PC was a Pentium 60 Gateway 200 PC (Gateway 2000, Inc.) and the whole system was put into a movable electronic cabinet for easy use.

References

American Society of Agricultural Engineers, *Moisture Measurement-Peanuts, ASAE Standards,* 44th ed., St. Joseph, MI., 1980.

American Society for Testing and Materials, Standard practice for determination of odor and taste thresholds by a forced-choice ascending concentration series method of limits, in *E679-91,* ASTM Committee E-18, Philadelphia, PA, 1991, 1.

American Society for Testing and Materials, Standard practice for referencing supra-threshold odor intensity, in *E544-75* (reapproved 1988), ASTM Committee E-18, Philadelphia, PA, 1988, 20.

Amrani, M. E. H., Persaud, K. C., and Payne, P. A., High frequency measurements of conducting polymers: development of a new technique for sensing volatile chemicals. *Measurement Science & Technology,* 6(10), 1500, 1995.

Bartlett, P. N., Elliott, J. M., and Gardner., J. W., Electronic noses and their application in the food industry, *Food Technol.,* 51, 44, 1997.

Berglund, B., Berglund, U., and Lindvall, T., Quality assurance in olfactometry, *Volatile Emissions from Livestock Farming and Sewage Operations*, Edited by V. C. Nielsen et al., Elsevier, New York, 1988.

Brailsford, A. D., Yussouff, M., and Logothetis, E. M., Steady state model of electrochemical gas sensors with multiple reactions, *Sensors and Actuators B*, 35–36, 392, 1996.

Brethour, J. R., Relationship of ultrasound speckle to marbling score in cattle, *J. Anim. Sci.*, 68, 2603, 1990.

Brynzari, V., Dmitriev, S., and Korotchenkov, G., Theory of the thin film gas sensor design, Int. Conf. Solid-State Sensors and Actuators, IEEE, 1997, 983.

Buehler, M. G., Technical support package on gas-sensor test chip, *NASA Tech. Brief.*, 21(8), item 115, 1997.

Cespedes, I. and Ophir, J., Elastography: experimental results from phantoms and tissues, *Ultrason. Imag.*, 13, 187, 1991.

Cespedes, I., Ophir, J., Ponnekanti, H., and Maklad, N., Elastography: elasticity imaging using ultrasound with application to muscle and breast *in vivo*, *Ultrason. Imag.*, 15, 73, 1993.

Chen, Y., Meat inspection system general description, University of Texas at Houston Medical School, Radiology Dept., 1996.

Christensen, D. A., *Ultrasonic Bioinstrumentation*, John Wiley & Sons, New York, 1988.

Collins, G. E. and Rose-Pehrsson, S. L., Chemiluminescent chemical sensors for inorganic and organic vapors, 8th Int. Conf. Solid-State Sensors and Actuators, and Eurosensors, IX, 1995.

Corcoran, P., The effects of signal conditioning and quantization upon gas and odour sensing system performance, *Sensors and Actuators B*, 18–19, 649, 1994.

Davide, F. A. M., Natale, C. D., and D'Amico, A., Self-organizing sensory maps in odour classification mimicking, *Biosens. Bioelectron.*, 10, 203, 1995.

Dickens, J. W., Slate, A. B., and Pattee, H. E., Equipment and procedures to measure peanut headspace volatiles, *Peanut Sci.*, 14, 96, 1987.

di Natale, C., Brunink, J. A. J., and Bungaro, F., Recognition of fish storage time by a metalloporphyrins-coated QMB sensor array, *Meas. Sci. Technol.*, 7, 1103, 1996.

Egashira, M., Functional design of semiconductor gas sensors for measurement of smell and freshness, Int. Conf. Solid-State Sensors and Actuators, IEEE, 1997, 1385.

Egashira, M. and Shimizu, Y., Odor sensing by semiconductor metal oxides, *Sensors and Actuators B*, 13–14, 443, 1993.

Eklov, T., Martensson, P., and Lundstrom, I., Enhanced selectivity of MOSFET gas sensors by systematical analysis of transient parameters, *Anal. Chim. Acta*, 353, 291, 1997.

Engen, T., *The Perception of Odors*, Academic Press, New York, 1982.

Ergun, D. L., Peppler, W. W., Dobbins, III, J. T., Zink, F. E., Kruger, D. G., Kelcz, F., de Bruijn, F. J., Nellers, E. B., Wang, Y., Althof, R. J., and Wind, M. G. J., Dual energy computed radiography: improvements in processing. Medical imaging: image processing, M. H. Loew, Ed., *Proc. SPIE*, 2167, 663, 1994.

Freund, M. S. and Lewis, N. S., Electronic noses made from conductive polymetric films. *NASA Tech. Brief*, 21(7), item 103, 1997.

Gardner, J. W., Craven, M., Dow, C., and Hines, E. L., The prediction of bacterial type and culture growth phase by an electronic nose with a multi-layer perceptron network, *Meas. Sci. Technol.* 9, 120, 1998.

Hanaki, S., Nakamoto, T., and Moriizumi, T., Artificial odor recognition system using neural network for estimating sensory quantities of blended fragrance, *Sensors and Actuators A,* 57, 65, 1996.

Henning, R., The hull-scrape method to determine when to dig peanuts, *The Peanut Farmer,* Aug., 11, 1983.

Huang, Y., Lacey, R. E., and Whittaker, A. D., Neural network prediction modeling based on ultrasonic elastograms for meat quality evaluation, *Trans. ASAE,* 41(4), 1173, 1998.

Huang, Y., Lacey, R. E., Whittaker, A. D., Miller, R. K., Moore, L., and Ophir, J., Wavelet textural features from ultrasonic elastograms for meat quality prediction, *Trans. ASAE,* 40(6), 1741, 1997.

HunterLab, Communicating color by numbers, company brochure, Hunter Associates Laboratory, Inc., Reston, VA, 1995.

Ishida, H., Nakamoto, T., and Moriizumi, T., Study of odor compass, IEEE/SICE/RSJ Int. Conf. Multisensor Fusion and Integration for Intelligence Systems, IEEE, 1996, 222.

Lacey, R. E. and Osborn, G. S., Applications of electronic noses in measuring biological systems, ASAE paper No. 98-6116, St. Joseph, MI, 1998.

Lawless, H. T., Olfactory psychphysics, in *Tasting and Smelling,* 2nd ed., Beauchamp, G. K., and Bartoshuk, L., Eds., Academic Press, San Diego, CA, 1997, 125.

Llobet, E., Brezmes, J., Vilanova, X., Fondevila, L., and Correig, X., Quantative vapor analysis using the transient response of non-selective thick-film tin oxide gas sensors, Transducers '97, Int. Conf. Solid-State Sensors and Actuators, IEEE, 1997, 971.

Lozano, M. S. R., Ultrasonic elastography to evaluate beef and pork quality, Ph.D. Dissertation, Texas A&M University, College Station, TX, 1995.

Matsuno, G., Yamazaki, D., Ogita, E., Mikuriya, K., and Ueda, T., A quartz crystal microbalance type odor sensor using PVC blended lipid membrane, *IEEE Trans. Instrum. Meas.* 44(3), 739, 1995.

Miasik, J. J., Hooper, A., and Tofield, B. C., Conducting polymer gas sensors, *J. Chem. Soc., Faraday Trans. 1,* 82, 1117, 1986.

Moore, L. L., Ultrasonic elastography to predict beef tenderness, M.S. thesis, Texas A&M University, College Station, TX, 1996.

Moriizumi, T., Saitou, A., and Nomura, T., Multi-channel SAW chemical sensor using 90 MHz SAW resonator and partial casting molecular films, *IEEE Ultrason. Symp.* 499, 1994.

Moy, L. and Collins, M., Electronic noses and artificial neural networks, *Am. Lab.,* Vol. 28 (February), 22, 1996.

Nakata, S., Akakabe, S., Nakasuji, M., and Yoshikawa, K., Gas sensing based on a nonlinear response: discrimination between hydrocarbons and quantification of individual components in a gas mixture, *Anal. Chem.,* 68(13), 2067, 1996.

Nanto, H., Sokooshi, H., and Kawai, T., Aluminum-doped ZnO thin film gas sensor capable of detecting freshness of sea foods, *Sensors and Actuators B,* 13–14, 715, 1993.

Nayak, M. S., Dwivedi, R., and Srivastava, S. K., Transformed cluster analysis: an approach to the identification of gases/odours using an integrated gas-sensor array, *Sensors and Actuators B,* 12: 103, 1993.

Ophir, J., Cespedes, I., Ponnekanti, H., Yazdi, Y., and Li, X., Elastography: a quantitative method for imaging the elasticity of biological tissues, *Ultrason. Imag.,* 13, 111, 1991.

Ophir, J., Miller, R. K., Ponnekanti, H., Cespedes, I., and Whittaker, A. D., Elastography of beef muscle, *Meat Sci.*, 36, 239, 1994.

Osborn, G. S., Young, J. H., and Singleton, J. A., Measuring the kinetics of acetaldehyde, ethanol, and ethyl acetate within peanut kernels during high temperature drying, *Trans. ASAE*, 39(3), 1039, 1995.

Osborn, G. S., Lacey, R. E., and Singleton, J. A., A method to detect peanut off-flavors using an electronic nose, *Trans. ASAE*, in press.

Park, B., Non-invasive, objective measurement of intramuscular fat in beef through ultrasonic A-mode and frequency analysis, Ph.D. dissertation, Texas A&M University, College Station, TX, 1991.

Park, B. and Whittaker, A. D., Ultrasonic frequency analysis for beef quality grading, ASAE paper No. 91-6573, St. Joseph, MI, 1991.

Park, B. and Whittaker, A. D., Determination of beef marbling score using ultrasound A-scan, ASAE paper No. 90-6058, St. Joseph, MI, 1990.

Persaud, K. C., DeSimone, J. A., and Heck, G. L., Physiology and psychophysics in taste and smell, in *Sensors and Sensory Systems for Advanced Robots*, Vol. F43, Dario, P., Ed., Springer-Verlag, Berlin, 1988, 49.

Ping, W. and Jun, X., A novel method combined neural network with fuzzy logic for odour recognition, *Meas. Sci. Technol.* 7, 1707, 1996.

Pisanelli, A. M., Qutob, A. A., Travers, P., Szyszko, S., and Persaud, K. C., Applications of multi array polymer sensors to food industry, *Life Chem. Rep.*, 11, 303, 1994.

Ponnekanti, H., Ophir, J., and Cespedes, I. Axial stress distributions between coaxial compressors in elastography: an analytical model, *Ultrasoun. Med. Biol.*, 18(8), 667, 1992.

Ruan, R., Schmidt, S., and Litchfield, J. B., Nondestructive measurement of transient moisture profiles and the moisture diffusion coefficient in potato during drying and absorption by NMR imaging, *J. Food Process. Eng.*, 14, 297, 1991.

Sayeed, M. S., Whittaker, A. D., and Kehtarnavaz, N. D., Snack quality evaluation method based on image features and neural network prediction, *Trans. ASAE*, 38(4), 1239, 1995.

Scott, A. O., *Biosensors for Food Analysis*, Royal Society of Chemistry, Cambridge, UK, 1998.

Simon, J. E., Hetzroni, A., Bordelon, B., Miles, G. E., and Charles, D. J., Electronic sensing of aromatic volatiles for quality sorting of blueberries, *J. Food Sci.* 61(5), 967, 972, 1996.

Singleton, J. A. and Pattee, H. E., Effect of chilling injury on windrowed peanuts, *Peanut Sci.*, 16, 51, 1989.

Singleton, J. A. and Pattee, H. E., A preconcentration and subsequent gas liquid chromatographic analysis method for trace volatiles, *J. Am. Oil Chem. Soc.*, 57, 405, 1980.

Strassburger, K. J., Electronic nose evaluation in the flavor industry; it really works!, *Food Test. Anal.*, 2, 22, 1996.

Sundgren, H., Winquist, F., and Lindstrom, I., Artificial neural networks and statistical pattern recognition improve MOSFET gas sensor array calibration, *IEEE*, 574, 1991.

Taylor, R. F., Applications of biosensors in the food processing industry, Proc. Food Process. Automa. Conf., ASAE, St. Joseph, MI, 1990, 156.

Thane, B. R., Prediction of intramuscular fat in live and slaughtered beef animals through processing of ultrasonic images, M.S. thesis, Texas A&M University, College Station, TX, 1992.

U. S. Department of Agriculture, *Official United States Standards for Grades of Carcass Beef*, 1997.

Vernat-Rossi, V., Garcia, C., Talon, R., Denoyer, C., and Berdague, J. L., Rapid discrimination of meat products and bacterial strains using semiconductor gas sensors, *Sensors and Actuators B*, 37, 43, 1996.

Wagner, G. and Guilbault, G. G., *Food Biosensor Analysis*, Marcel Dekker, NY, 1994.

White, J., Kauer, J. S., Dickinson, T. A., and Walt, D. R., Rapid analyte recognition in a device based on optical sensors and the olfactory system, *Anal. Chem.* 68(13), 2191, 1996.

Whittaker, A. D., Park, B., Thane, B. R., Miller, R. K., and Savell, J. W., Principles of ultrasound and measurement of intramuscular fat, *J. Anim. Sci.*, 70, 942, 1992.

Wide, P., Winquist, F., and Driankov, D., An air-quality sensor system with fuzzy classification. *Meas. Sci. Technol.*, 8, 138, 1997.

Williams, A. G., Physical calibration of olfactometric measurements, in *Odour Prevention and Control of Organic Sludge and Livestock Farming*, Edited by V. C. Nielsen et al., Elsevier, New York, 1986.

Williams, E. J. and Drexler, J. S., A non-destructive method for determining peanut pod maturity, *Peanut Sci.*, 8, 134, 1981.

Winquist, F., Hornsten, E. G., Sundgren, H., and Lundstrom, I., Performance of an electronic nose for quality estimation of ground meat, *Meas. Sci. Technol.*, 4, 1493, 1993.

Yea, B., Konihi, R., Osaki, T., and Sugahara, K., The discrimination of many kinds of odor species using fuzzy reasoning and neural networks, *Sensors and Actuator A*, 45, 159, 1994.

Yim, H. S., Kibbey, C., Ma, S. C., Kliza, D. M., Liu, D., Park, S. B., Torre, C. E., and Meyerhoff, M. E. Polymer membrane-based ion-, gas- and bio-selective protensensors, *Biosens. Bioelectronics.*, 8, 1, 1993.

Zeng, X., Ruan, R., Fulcher, R., and Chen, P., Evaluation of soybean seedcoat cracking during drying, *Drying Technol.*, 14 (7–8), 1595, 1996.

chapter three

Data analysis

Analysis of acquired data is an important step in the process of food quality quantization. Data analysis can help explain the process it concerns. Also, the analysis is beneficial for determining whether the available data is usable to extract the information to fulfill the goals in problem solving. In general, there are two kinds of data analysis. One is the analysis for static relationships, called static analysis. For example, in food quality classification and prediction, the functions between input and output variables are usually static. That is to say, such input and output relationships may not vary with time. The other kind of data analysis, dynamic analysis, seeks dynamic relationships within the process. This second kind is usually needed for food quality process control because in food process modeling and control, the relationships that are mainly dealt with are dynamic. This means that these relationships change with time. In this chapter, these two kinds of data analysis, static and dynamic, will be discussed with practical examples in food engineering.

Images are an important data type in food engineering applications. Image analysis is conducted through image processing. In this chapter, image processing will be discussed for the purpose of image analysis. Through image processing, image pixel values are converted into numerical data as the input parameters to modeling systems.

3.1 Data preprocessing

Before data analysis, data preprocessing is necessary to remove "noise" from the data to let analysis and modeling tools work on "clean" data covering similar ranges.

In general, data preprocessing involves scaling all input and output data from a process to a certain range and the reduction of dimensionality. Many tools of analysis and modeling work better with data within similar ranges, so it is generally useful to scale raw input and output data to a common mean and range. The scaling methods are usually used to scale the input and output data to a mean of zero and a standard deviation of one or to a mean of zero with a range of plus or minus one.

Assume process input and output observations are u_k and y_k, then the data with zero mean and one standard deviation are

$$\tilde{u}_k = \frac{u_k - \bar{u}}{s_u}$$

$$\tilde{y}_k = \frac{y_k - \bar{y}}{s_y} \qquad (k = 1, \ldots, N) \tag{3.1}$$

where $\bar{u}$ and $\bar{y}$ are the means of raw input and output data, respectively; s_u and s_y are the standard deviations of raw input and output data, respectively; k is a data sample sequential number; and N is the sample size or the number of samples.

In data preprocessing, a good example is that, for artificial neural network modeling (discussed further in later chapters), all input and output data should be scaled to the range of the processing units of the networks. The most popular transfer function of the network processing units is the sigmoidal function

$$S(x) = \frac{1}{1 + e^{-x}}$$

where x is the input of one of the processing units in the network. The range of the function is [0, 1]. So, the input and output data are often scaled as follows to fall into the range

$$\tilde{u}(k) = \frac{u(k) - u_{\min}}{u_{\max} - u_{\min}}$$

$$\tilde{y}(k) = \frac{y(k) - y_{\min}}{y_{\max} - y_{\min}} \tag{3.2}$$

where $u_{\max}$ and $u_{\min}$ are the maximum and minimum values of the raw input data, respectively, and $y_{\max}$ and $y_{\min}$ are the maximum and minimum values of the raw output data, respectively. This scaling method scales the data into the range of [0, 1]. If any other ranges are required, only simple arithmetic is needed on the scaled data $\tilde{u}(k)$ and $\tilde{y}(k)$. For example, if a range of [−1, 1] is required, then $2\tilde{u}(k) - 1$ and $2\tilde{y}(k) - 1$ produce the data in this range that can actually be used to construct a different nonlinear transfer function of processing units of the networks

$$S(x) = \frac{1 - e^{-x}}{1 + e^{-x}}$$

For dynamic systems, input and output data often contain constant or low frequency components. No model identification methods can remove the negative impact of these components on the modeling accuracy. Also in many cases, high frequency components in the data may not be beneficial for model identification. In general data analysis and modeling of dynamic systems, the input and output data need to be preprocessed to zero mean to eliminate high frequency components. This may significantly improve the accuracy of model identification.

For many years, transform theory has played an important role in data preprocessing and analysis. There are various types of transforms, but our emphasis is on the methods of Fourier and wavelet transforms because of their wide range applications in data preprocessing and analysis problems.

Fourier transform is a classic tool in signal processing and analysis. Given a signal $f(x)$, the one-dimensional discrete Fourier transform of it can be computed by

$$F(k) = \frac{1}{N} \sum_{i=0}^{N-1} f(i) e^{-j2\pi ki/N} \tag{3.3}$$

for $k = 1, 2, \ldots, N - 1$ and $j = \sqrt{-1}$. The signal $f(x)$ can be reconstructed by

$$f(x) = \sum_{i=0}^{N-1} F(i) e^{-j2\pi xi/N} \tag{3.4}$$

for $x = 1, 2, \ldots, N - 1$.

If Fourier transform is expressed as follows

$$\Im(f(x)) = F(k)$$

then,

$$\Im(f(x - x_0)) = F(k) e^{-j2\pi kx_0/N} \tag{3.5}$$

This is the translation property of the Fourier transform. In spectral analysis, the magnitude of the Fourier transform of $f(x)$ is displayed as

$$|\Im(f(x))| = |F(k)| \tag{3.6}$$

Note from Eq. (3.5) that a shift in the signal $f(x)$ does not affect the magnitude of its Fourier transform

$$\left| F(k) e^{-j2\pi kx_0/N} \right| = |F(k)|$$

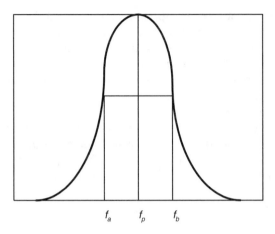

$$f_a \qquad f_p \qquad f_b$$

Figure 3.1 Spectrum power curve for ultrasonic transducer. (Adapted from Park, 1991. With permission.)

In food quality inspection, frequency analysis techniques are useful to reveal characteristics of materials. Frequency analysis is based on discrete Fourier transformation of signal data from experiments, such as ultrasonic A-mode experiments. Then, spectral analysis will indicate which frequencies are most significant. Let us study a little bit more detail about frequency analysis for ultrasonic A-mode experiments. Figure 3.1 shows an ideal spectrum curve for a homogeneous ultrasonic signal. This curve is in a Gaussian, or normal, shape. The spectrum is symmetric about the peak frequency, f_p; thus, there is no skewness. The frequency half-power points, f_a and f_b, in this case occur equidistant to either side of the peak. The central frequency can be calculated as

$$f_c = \frac{1}{2}(f_a + f_b) \tag{3.7}$$

For a symmetric spectrum, f_p, peak-power (resonant) frequency is equal to f_c, the central frequency. The percentage bandwidth, B^*, expresses the broadness of the curve as

$$B^* = \frac{f_b - f_a}{f_c} \times 100 \tag{3.8}$$

The skewness, f_{sk}, can be represented as

$$f_{sk} = \frac{f_p - f_a}{f_b - f_p} \tag{3.9}$$

Local maxima is another parameter which is useful in spectral analysis. It describes the multiple peaks of the Fourier spectrum from an ultrasonic signal.

Wavelet transform presents a breakthrough in signal processing and analysis. In comparison to the Fourier transform, wavelet basis functions are local both in frequency and time while Fourier basis functions are local only in frequency. Wavelets are "small" waves that should integrate to zero around the x axis. They localize functions well, and a "mother" wavelet can be translated and dilated into a series of wavelet basis functions. The basic idea of wavelets can be traced back to very early in the century. However, the development of the construction of compactly supported orthonormal wavelets (Daubechies, 1988) and the wavelet-based multiresolution analysis (Mallat, 1989) have resulted in extensive research and applications of wavelets in recent years.

In practice, the wavelet analysis can be used to transform the raw signals, process the transformed signals, and transform the processed results inversely to get the processed, reconstructed signals. Given a signal $f(x)$, a scaling function $\Phi(x)$, and a wavelet function $\Psi(x)$, the one-dimensional discrete orthogonal wavelet transform of $f(x)$ along the dyadic scales can be computed by Mallat's recursive algorithm

$$f^{(m)}(k) = \sum_i h(i - 2k) f^{(m+1)}(i) \tag{3.10}$$

$$d^{(m)}(k) = \sum_i g(i - 2k) f^{(m+1)}(i) \tag{3.11}$$

where $f^{(m)}(k)$ is the smoothed signal of $f(x)$ at the resolution m and the sampling point k, $d^{(m)}(k)$ is the detail signal of $f(x)$ at the resolution m and sampling point k, $h(i)$ is the impulse response of a unique low-pass FIR (finite impulse response) filter associated with $\Phi(x)$ at the sampling point k, $g(i)$ and is the impulse response of a unique FIR filter associated with $\Phi(x)$ and $\Psi(x)$ at the sampling point k.

This computation is a convolution process followed by $j/2$ subsampling at one-half rate. Here $j = N, N - 1, \ldots, 1$ and $N = 2^M$, where M is the highest level of the resolution of the signal, is the number of signal samplings. In the preceding equations, $f^{(m)}$ is the smoothed signal at scale 2^m while $d^{(m)}$ is the detail signal that is present in $f^{(m)}$ but lost in $f^{(m+1)}$ at scale 2^{m+1}.

An image is a two-dimensional data array. The concepts and methods described previously are all extendable to the two-dimensional case. Especially, we will see such an extension of wavelet transform to two dimensions in Section 3.3.2. These two-dimensional tools are very useful in image processing and analysis.

Image preprocessing is important for human perception and subsequent analysis. A poorly preprocessed image will be less understood by a human or computer analyzer. It is critical to remove the noises, adhere to the image, and enhance the region that we are concerned with in order to ensure the performance of an imaging system.

Images are subject to various types of noises. These noises may degrade the quality of an image and, hence, this image may not provide enough information. In order to improve the quality of an image, operations need to be performed on it to remove or decrease degradations suffered by the image in its acquisition, storage, or display. Through such preprocessing, the appearance of the image should be improved for human perception or subsequent analysis.

Image enhancement techniques are important in image preprocessing. The purpose of image enhancement is to process an image to create a more suitable one than the original for a specific application. Gonzalez and Woods (1992) explained that the word *specific* is important because it establishes at the outset that the enhancement techniques are very much problem-oriented. Thus, for example, a method that is quite useful for enhancing x-ray images may not necessarily be the best approach for enhancing pictures of apples taken by a digital camera.

Image enhancement techniques can be divided into two categories: spatial domain methods and frequency domain methods. Spatial domain refers to the image plane itself. Approaches in this category are based on direct manipulation of pixels in an image which includes point processing and spatial filtering (smoothing filtering and sharpening filtering) (Gonzalez and Woods, 1992). Frequency domain processing techniques are based on modifying the Fourier transform of an image which involves the design of lowpass, highpass, and homomorphic filters (Gonzalez and Woods, 1992).

3.2   Data analysis

Data analysis is necessary before model building and controller design. Data analysis captures the relationships between inputs and outputs. The analysis is basically performed by the methods of correlation analysis in classical statistics. The correlation analysis is a tool to show qualitatively the connection between inputs and outputs. It is used to determine the degree of connection between variables but does not account for the causality. The outcome of data analysis strongly assists accurate process modeling and effective process control. Next, static and dynamic data analysis will be discussed based on correlation analysis.

3.2.1   Static data analysis

Static data analysis is based on the cross-correlation analysis between variables. The cross-correlation function is used to detect the relationships between different variables at the same time instant. For example, in food quality quantization, the cross correlation between input variables, such as features of electronic measurement data on food samples, and output variables, such as food sensory attributes, is needed.

The correlation between variables can be simple or multiple. The simple correlation refers to the connection between two variables while the multiple correlation refers to the connection between three or more variables. The simple correlation is the basis of correlation theory. It will be explained in this section. The multiple correlation is closely related to multivariate regression. It can be extended based on simple correlation. Interested readers can refer to books on this topic.

Scatter plot is a basic graphic tool for correlation analysis. It plots one variable vs. the other in a two-dimensional co-ordinate system. The correlation between two variables can be positive, negative, or uncorrelated. If two variables change in the same direction, that is, they increase and decrease together, they are said to be positively correlated. If two variables change in the opposite direction, that is, when x increases, y decreases, or when x decreases, y increases, they are said to be negatively correlated. If two variables have no connection in change at all, they are said to be uncorrelated or zero correlated. In addition, the correlation can be linear or nonlinear. Linear correlation means that all data points in the scatter plot gather around a straight line. Nonlinear correlation means that all data points in the scatter plot form around a curve. Figure 3.2 shows the positive correlation in linear and nonlinear cases. Figure 3.3 shows the negative correlation in linear and nonlinear cases. Figure 3.4 shows the situation at which data points scatter in the whole x–y plane without a function pattern.

Through the preceding description, the type and strength of correlation between two variables can be understood by direct observation of the scatter plot. If data points are closely around a straight line or a curve, they are strongly correlated. If data points scatter from a straight line or a curve, they are weakly correlated or even uncorrelated. However, a scatter plot only gives a rough profile of the correlation between two variables. For precise measurement of the correlation between two variables, a statistic of correlation coefficient is used. Conventionally, the symbol ρ is used to represent the correlation coefficient of the population of x and y, written as ρ_{xy} while the symbol r is used to represent the estimated sample correlation coefficient, written as r_{xy}. The equation for calculation of a sample correlation coefficient is as follows:

$$r_{xy} = \frac{\sum\limits_{i=1}^{N}(x_i - \bar{x})(y_i - \bar{y})}{\sqrt{\sum\limits_{i=1}^{N}(x_i - \bar{x})^2}\sqrt{\sum\limits_{i=1}^{N}(y_i - \bar{y})^2}} \qquad (3.12)$$

where N is the number of samples. In practice, the correlation coefficient of the population is unknown while the sample correlation coefficient can always be calculated to estimate the population correlation coefficient. Eq. (3.12) is widely used in practical data analysis. The following examples present the use of the correlation coefficient r_{xy} for static data analysis in the process of food quality quantization.

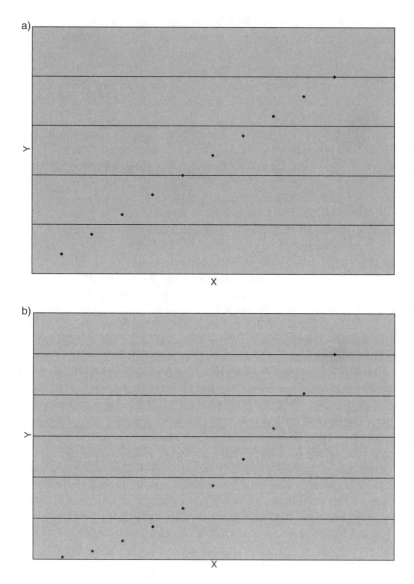

Figure 3.2 Scatter plots of (a) linear and (b) nonlinear positively correlated relationships.

3.2.1.1   Example: Ultrasonic A-mode signal analysis for beef grading
One of the primary factors in determining beef grades is the amount of intramuscular fat or marbling. Table 3.1 shows the parameters used by human graders in determining marbling levels involves visual inspection of a cross-sectional area of the ribeye steak. In the study, the specimens were selected ranging from practically devoid to abundant marbling score as assigned by human graders. The fat concentration of these specimens by

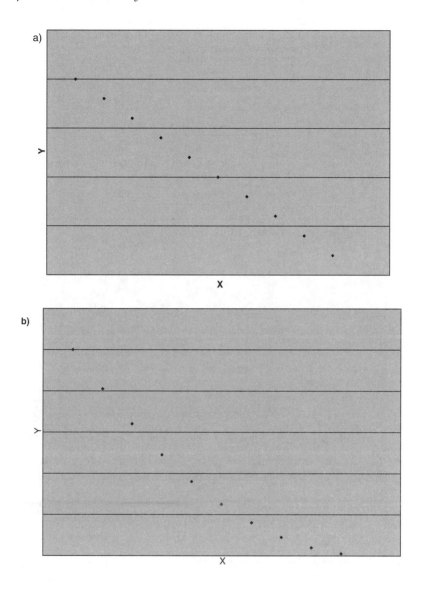

Figure 3.3 Scatter plots of (a) linear and (b) nonlinear negatively correlated relationships.

biochemical analysis (ether extraction) was different. Actually, the cross-correlation coefficient between the visual marbling score and fat concentration was 0.7.

At the initial stage to quantify the beef grade, Park (1991) performed correlation analysis between the ultrasonic signals and the marbling levels designed by human graders. The ultrasonic speed and attenuation were measured with the ultrasonic analyzer. The ultrasonic longitudinal speed

Table 3.1 Mean Ether Extractable Fat of Beef Longissimus Steaks
Stratified According to Marbling Level*

Marbling Level	Marbling Score	Fat(%) Mean	Std. Dev.
Moderately abundant	800–899	10.42	2.16
Slightly abundant	700–799	8.56	1.60
Moderate	600–699	7.34	1.50
Modest	500–599	5.97	1.15
Small	400–499	4.99	1.10
Slight	300–399	3.43	0.89
Traces	200–299	2.48	0.59
Practically devoid	100–199	1.77	1.12

* Adapted from Park (1991). With permission.

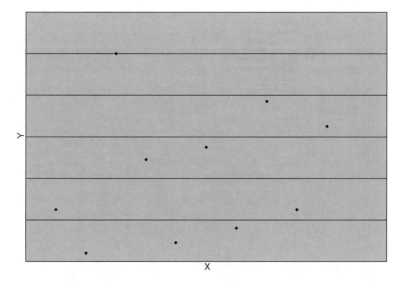

Figure 3.4 Scatter plot of noncorrelated data.

showed a gradual decrease with the increase of the fat concentration in the beef samples. Figure 3.5 shows that the fat concentration and the ultrasonic longitudinal speed were negatively correlated. The value of the correlation was −0.82, which was significantly higher than −0.72, the one between the visual marbling score and longitudinal speed.

Ultrasonic attenuation gradually increased as fat concentration increased, that is, they were positively correlated with each other. The cross-correlation coefficients between them increased at 0.24, 0.36, and 0.47 when the probe frequencies were 1 MHz, 2.25 MHz, and 5 HMz, respectively. This indicated that the attenuation was more sensitive at higher frequencies than at lower

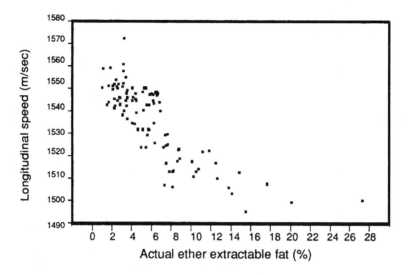

Figure 3.5 Fat concentration and ultrasonic longitudinal speed relationship. (From Park, 1991. With permission.)

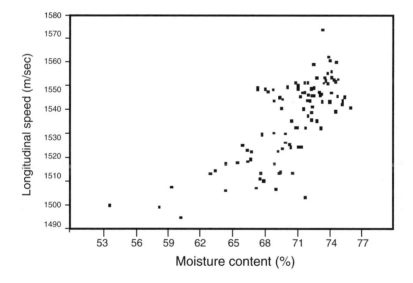

Figure 3.6 Fat concentration and moisture content relationship. (From Park, 1991. With permission.)

probe frequencies. The main reason for the low correlation coefficients was that the measurement of the attenuated signal was not precise enough.

Figure 3.6 shows that the fat concentration of the beef samples mono-tonically decreased as the moisture content of the samples increased. The cross-correlation coefficient between them was −0.92.

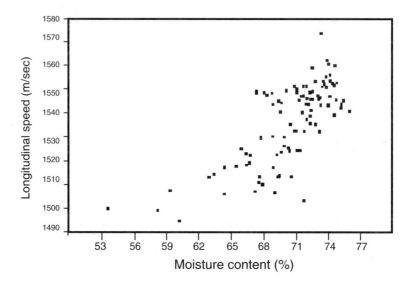

Figure 3.7 Moisture content and ultrasonic longitudinal speed relationship. (From Park, 1991. With permission.)

The ultrasonic speed and attenuation coefficients were correlated with the moisture content (% wet weight basis) in the beef samples. As shown in Figure 3.7, the ultrasonic longitudinal speed monotonically increased as the moisture content increased with the cross-correlation coefficient of 0.74. The attenuation coefficient appeared to remain the same or decrease as the moisture content of beef samples increased, but no significant trends appeared for the attenuation coefficient to be correlated with the moisture content.

The preceding correlation analysis indicated initially that ultrasonic longitudinal speed could be an important factor for the estimation of the intramuscular fat in the time domain.

With frequency analysis, it is possible to provide sufficient information about marbling. In frequency analysis experiments, broadband, highly damped probes send a wide range of frequencies into the sample. Spectral analysis then indicates which frequencies are most affected by the amount of marbling found in the longissimus muscle area. Following the Fourier transform of a RF signal, seven parameters in the frequency domain can be recorded as shown in Table 3.2 which were described earlier in Section 3.1.

Frequency analysis results should designate a particular central frequency that experiences the most significant spectrum change between different marbled meat samples. With a probe having this central frequency, a sample with lower marbling should be indicated by a large increase in the echo height compared with a sample with higher marbling. With this anticipated result, changes in echo amplitude (attenuation) may be more useful as the indicating parameter for detecting intramuscular fat (marbling) in the longissimus muscle area.

Table 3.2 Ultrasonic Parameters in the Frequency Domain*

Parameter	Description
f_a	Frequency at lower one-half power point in the power spectrum
f_b	Frequency at higher one-half power point in the power spectrum
f_p	Peak pulse frequency of the power spectrum
f_c	Central pulse frequency of the power spectrum
B*	Bandwidth
f_{sk}	Skewness of the power spectrum
Lm	Number of local maxima

* Adapted from Park (1991). With permission.

Table 3.3 Pearson Correlation Coefficients between Fat Concentration and Parameters in the Frequency Domain[†]

	Fat Concentration					
Probe	Longitudinal (MHz)			Shear (MHz)		
Parameter	1	2.25	5	1	2.25	5
N (number of samples)	60	97	60	61	96	60
f_a	0.44	−0.15	0.12	0.13	0.01	0.39
f_b	0.22	−0.25	0.04	0.12	−0.02	0.13
f_p	0.43	−0.20	0.08	0.06	0.02	0.27
f_c	0.36	−0.20	−0.08	−0.36	−0.01	0.29
B*	−0.45*	−0.10	−0.35	0.34	−0.05	−0.36
f_{sk}	0.07	−0.07	−0.06	0.13	0.24	−0.11
Lm	−0.10	0.59*	0.68*	0.68*	0.89**	0.76*

* Significant parameters within same probe.

** The most significant parameter between different probes.

[†] Adapted from Park (1991). With permission.

Park (1991) initially investigated the seven parameters of the Fourier spectra to determine which ones are more affected by marbling score through cross-correlation coefficients. The number of local maxima was the most correlated parameter with fat concentration for each ultrasonic probe except the 1 MHz longitudinal probe. In that case, the bandwidth was more correlated than other parameters as shown in Table 3.3.

Among the different probes, the parameter values for the shear probes were more highly correlated than those for the longitudinal probes. For example, the correlation coefficient between number of local maxima and fat concentration for longitudinal probes were 0.59 in 2.25 MHz and 0.68 in 5 MHz; whereas, those values for shear probes were 0.89 in 2.25 MHz and 0.76 in 5 MHz, respectively.

Table 3.4 Pearson Correlation Coefficients between Moisture Content and Parameters in the Frequency Domain[†]

| Probe Parameter | Moisture Content | | |
| | Longitudinal (MHz) | | |
	1	2.25	5
N (number of samples)	60	97	60
f_a	−0.41	0.24	−0.10
f_b	−0.04	0.28	−0.01
f_p	−0.38	0.25	−0.07
f_c	−0.25	0.26	−0.06
B^*	0.56*	−0.08	0.31
f_{sk}	−0.26	−0.03	0.01
Lm	0.06	−0.36*	−0.72**

* Significant parameters within same probe.

** The most significant parameter between different probes.

[†] Adapted from Park (1991). With permission.

This result shows that the ultrasonic shear probe is more sensitive to intramuscular fat concentration than the longitudinal probe. As a result of parameter analysis, the number of local maxima in the 2.25 MHz shear probe was the most significant parameter for predicting intramuscular fat concentration. The correlation coefficient between fat concentration and the number of local maxima was 0.89 in the 2.25 MHz shear probe.

Also, the number of local maxima was the most correlated parameter with moisture content for each ultrasonic probe except the 1 MHz longitudinal probe, in which the correlation of bandwidth was higher than any other parameter. As shown in Table 3.4, the correlation coefficient between the number of local maxima and the moisture content for longitudinal probes were −0.36 in 2.25 MHz and −0.72 in 5 MHz. This result shows that the ultrasonic longitudinal probe also is sensitive to moisture content. As a result of the parameter analysis, the number of local maxima in the 5 MHz longitudinal probe was the most significant parameter for predicting moisture content, of which the correlation coefficient between moisture content and the number of local maxima was −0.72.

According to experimental results, increasing intramuscular fat concentration corresponds to an increase of the number of local maxima, and increasing moisture content corresponds to a decrease of the number of local maxima. In this study, the number of local maxima was counted manually from the Fourier spectra of the ultrasonic signal. Figure 3.8 shows an example of typical discontinuity in the frequency spectra, from which the number of local maxima in a practically devoid marbling meat sample was three.

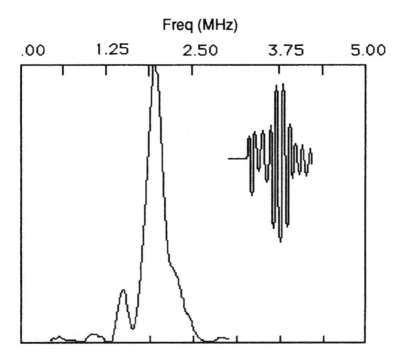

Figure 3.8 Fourier spectrum of 1.111 percent fat (practically devoid) with a 2.25 MHz shear probe. (From Park, 1991. With permission.)

3.2.1.2 Example: Electronic nose data analysis for detection of peanut off-flavors

Temperature and humidity cause certain effects to the performance in electronic nose applications. Temperature variation during testing results in two compounding errors. Temperature affects the partition coefficient between the gaseous phase and the absorbed phase for each volatile molecule. This alters the amount of the molecule absorbed to the sensor and, thus, the sensor response. Additionally, many gas sensors used in electronic noses are resistive devices whose resistance is a function of temperature as well as absorbed molecules. Temperature effects from self-heating in signal conditioning circuits must also be taken into account.

Like temperature, humidity creates two potentially confounding effects in electronic nose applications. First, semiconducting polymer sensors are very reactive to humidity, and small changes in water vapor can overshadow them. Figure 3.9 shows the response of a commercial electronic nose (Neotronics, Flowery Branch, GA) to water vapor pressure for a series of ground peanut samples. The samples were at the same moisture content, and vapor pressure effects were thought to be caused by absorption kinetics by the sensors. Although only 1 of the 12 sensors is shown, all sensors in the instrument had

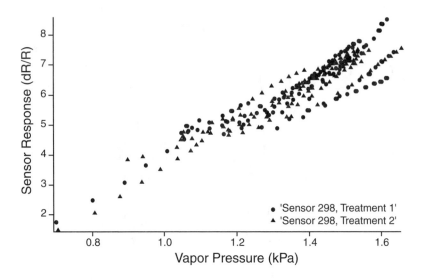

Figure 3.9 Response of a single sensor in the Neotronics electronic nose to vapor pressure for ground peanut samples. (From Lacey and Osborn, 1998. With permission.)

a similar response. The dependence on vapor pressure masked any variability from the sample treatments.

The AromaScan electronic nose was set up for the experiments to determine if significant differences could be detected between peanuts cured at high temperatures containing off-flavors and peanuts cured at low temperatures not containing off-flavors (Osborn et al., in press). For the experiments, sensor readings were averaged over a time period of 5 s. This procedure produced a total of 60 data points (300 s test length/5 s each data point) for each sensor. Data was collected in this manner in an attempt to determine the best sample time for separating the two curing treatments. There was also some indication that the sensor readings would not stabilize during the test so the kinetics behavior of the sensor output was needed. Figure 3.10 shows a sample of the data collected from a single sensor on the low temperature cured ground kernels for all 10 test replications. The data shown in Figure 3.11 is the data from the same sensor for the high temperature cured ground kernels. Note that the low temperature readings appear to be generally constant over time, while the high temperature readings appear to increase as the test progresses. Also note the 10 replications have differences in the "offset" value, but the slope appears to be consistent between replications. All sensors and tests are not shown owing to space limitations for this book. The two figures shown are representative of the trends found on most tests.

Sample means were compared using a *t* test between each sensor at each time for each level of sample destruction (whole pod, whole kernel, half kernel, and ground kernel). For example, data from sensor 11 for the low temperature cured ground kernels was compared to the sensor 11 data from the high temperature ground kernel test at each time step (0 through 300 s).

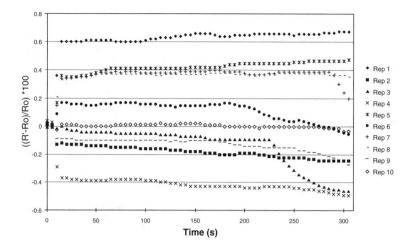

Figure 3.10 Raw AromaScan data from sensor 11, ground kernels, low temperature curing treatment for 10 replications. (From Osborn et al., in press. With permission.)

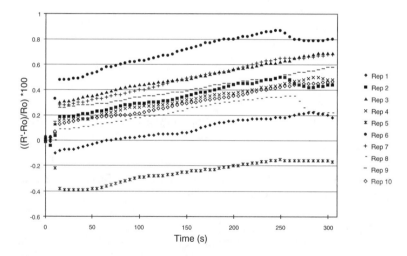

Figure 3.11 Raw AromaScan data from sensor 11, ground kernels, high temperature curing treatment for 10 replications. (From Osborn et al., in press. With permission.)

Raw value of the change in sensor resistance per initial sensor resistance $(\Delta R/R_0) \times 100$ was compared without any further data processing. For each *t* test (sample size 10, degree of freedom 9) equal variances between samples were assumed. Typical results for the *t* test are shown in Figure 3.12 for the sensors 8 through 15 for ground kernels. Note, the peak *t* value is attained at the end of the test. If a significance level of 0.05 is assumed, the *t* value for a 2 tailed *t* test must exceed 2.101 in order for the means of the sensor readings at each time step to be considered significantly different. Note in

Table 3.5 Test Results for All Data Collected[†]

	Whole Pods	Whole Kernels	Half Kernels	Ground Kernels
No. of sensors significant ($\alpha = 0.05$)	0/32	0/32	3/32	26/32
Highest t value (any sensor)	1.21	0.651	2.66	6.68
Time at peak t (s) (any sensor)	135	235	230	305

[†] Adapted from Osborn et al. (1998). With permission.

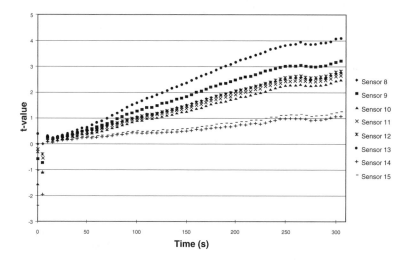

Figure 3.12 Raw AromaScan data t-test comparing sensors, ground kernels, and sensors 8 through 15. (From Osborn et al., in press. With permission.)

Figure 3.12 that six sensors are significantly different between curing treatments, and two are not significantly different. Table 3.5 summarizes the t test results for all the data collected.

In summary, t tests conducted on each sensor showed 26 sensors with significant differences between curing treatments for ground kernels. Only three sensors showed significant differences for whole kernels. The half kernels and whole pod tests were not significantly different between high and low temperature curing.

3.2.2   Dynamic data analysis

Dynamic data analysis is based on the autocorrelation analysis of the same variable and the cross-correlation analysis between different variables at different time instants and lags. The autocorrelation is used to detect the relationships in a variable itself at different time lags. For example, in modeling for food quality process control, autocorrelation analysis of the controlled variables, such as product quality indicating variables in color and moisture content, is needed to find out the most significant relationship of variables between current instant t and some previous instant $t - t_d$ to determine the order of the model.

In dynamic data analysis, the autocorrelation coefficients of process data sequences, or time series, are computed. Besides scatter plotting, a sample correlogram is a graphic technique in dynamic data analysis. Dynamic data sequences exhibit the characteristic of sequential correlation, that is, correlation over time. For example, Figure 3.13 shows two scatter plots of series $y_1(t)$ vs. $y_1(t - 1)$ and series $y_2(t)$ vs. $y_2(t - 1)$. From the plots, the series $y_1(t)$ appears to be high positively autocorrelated, and the series $y_2(t)$ appears not to be correlated.

In order to describe the sequential correlation numerically and graphically, Eq. (3.12) is extended to calculate sample autocorrelation coefficients. Unlike static data analysis, in dynamic data analysis the measurements are needed for the correlation of the data sequence with itself in certain time lags. Therefore, the following equation can be defined to evaluate the sample autocorrelation coefficient of lag l for the sequence $y(1), y(2),..., y(N)$

$$r_y(l) = \frac{\sum\limits_{t=1}^{N-l}(y(t) - \bar{y})(y(t + l) - \bar{y})}{\sum\limits_{t=1}^{N}(y(t) - \bar{y})^2}, \quad l < N \tag{3.13}$$

A plot of $r_y(l)$ vs. l for $l = 1, 2,..., l_{max}$ where l_{max} is a maximum time lag is called the correlogram of the data sequence.

There are some points that need to be noted in Eq. (4.4). $r_y(0) = 1$ and $l_{max} = N - 1$, which means that the maximum possible value of time lag l is $N - 1$. Also, $r_y(l) = r_y(-l)$.

For further dynamic input and output data sequences $u(1), u(2),..., u(N)$ and $y(1), y(2),..., y(N)$, the sample cross-correlation coefficient of lag l can be evaluated by the following equation

$$r_{uy}(l) = \begin{cases} \dfrac{\sum\limits_{t=1}^{N-l}(u(t) - \bar{u})(y(t + l) - \bar{y})}{\sqrt{\sum\limits_{t=1}^{N}(u(t) - \bar{u})^2 \sum\limits_{t=1}^{N}(y(t) - \bar{y})^2}} & l = 0,..., N - 1 \\ (r_{yu}(-l), & l = -(N - 1),..., 0) \end{cases} \tag{3.14}$$

It should be noted that $r_{uy}(l)$ is not symmetric about the point $l = 0$. A plot of $r_{uy}(l)$ vs. l for $l = -l_{max},..., l_{max}$ is called the cross correlogram of $u(t)$ and $y(t)$.

In the following example, you can see the use of correlation coefficients and correlograms in the data analysis for the dynamical characteristics of the snack food frying process.

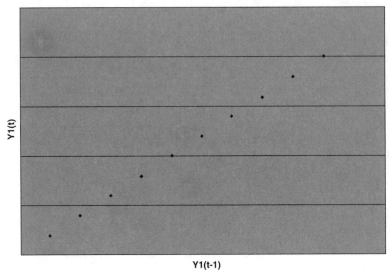

(a) High positively autocorrelated

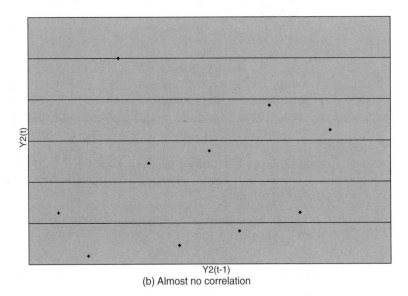

Y2(t-1)

(b) Almost no correlation

Figure 3.13 Scatter plots of series $y_1(t)$ vs. $y_1(t-1)$ and series $y_2(t)$ vs. $y_2(t-2)$.

3.2.2.1 *Example: Dynamic data analysis of the snack food frying process*

The input–output data of the snack food frying process were obtained according to the sampling procedure described earlier. In the operating ranges of the process in the experiments, the process input sequence, $u(t)$, needs to be able to keep stimulating process dynamics adequately so that the process output sequence, $y(t)$, can present process characteristics as much as possible. Several

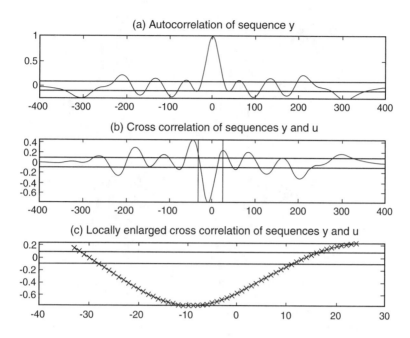

Figure 3.14 (a) Autocorrelogram of color $y(t)$. (b) Crosscorrelogram between inlet temperature $u(t)$ and color $y(t)$. (c) Local enlargement of (b). (From Huang and Whittaker, 1993. With permission.)

single-input single-output (SISO) and multiple-input multiple-output (MIMO) data sets were sampled for different products and purposes. Here, only a SISO set is demonstrated in order to explain the use of correlation analysis to reveal the process dynamics, while the correlation analysis can be similarly operated on other process input–output data of SISO and/or MIMO systems.

Specifically, the SISO data for demonstration were sampled every 10 s. The time lag between the process input, inlet temperature, and output, color, was measured around 100 s in the production line. The cross-correlation function between the inlet temperature and color was computed. Figures 3.14(b) and (c) show that the inlet temperature as the input sequence, $u(t)$, and the color as the output sequence, $y(t)$, have the highest correlation around 10, which is obviously caused by the time lag ($10 \times 10 = 100$) between the process input and output. Further, it is clear to see in Figure 3.15 that, overall, the color sequence, $y(t)$, has no clear linear relationship with the inlet temperature sequence, $u(t)$, but there exists a linear link between them. Owing to the 100 s time lag between the process input and output, when the lag parameter, d, is much smaller or much larger than 10, the relationship between $u(t - d)$ and $y(t)$ is not quite clear. Only when d is around 10, the relationship between $u(t - d)$ and $y(t)$ is clearly nonlinear where there is a linear component. Figure 3.16 shows the linear relationship between $y(t)$ and $y(t - i)$ ($i = 1, 2,...$), in which the smaller i is, the clearer the linear relationship between them. The autocorrelation function of $y(t)$ in Figure 3.14(a) verifies this relationship.

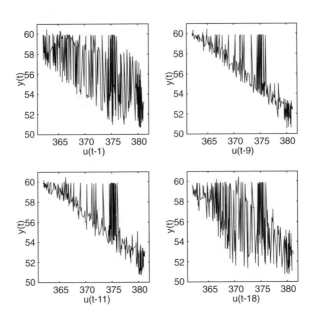

Figure 3.15 Scatter plot between inlet temperature $u(t - d)$ and color $y(t)$. (From Huang and Whittaker, 1993. With permission.)

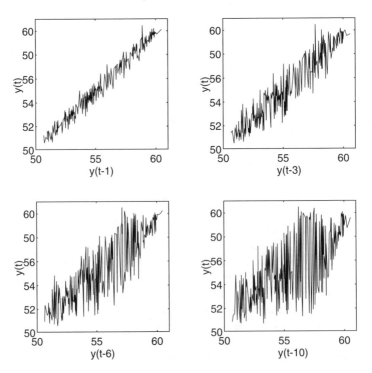

Figure 3.16 Scatter plot between color outputs in $y(t - i)$ and $y(t)$. (From Huang and Whittaker, 1993. With permission.)

3.3 Image processing

Imaging is an important way to quantify food quality. As described previously, there are several methods such as machine vision, medical imaging, and so on to acquire images of food samples. Once the images are available, the approaches to image processing are similar. Image preprocessing and image analysis are two major steps in image processing. Image preprocessing was discussed in Section 3.1 for image denoising and enhancement, and image processing is for extracting information from images and includes image segmentation and feature extraction, mainly in the process of food quality quantization.

3.3.1 Image segmentation

To segment an image is an important step in image processing and analysis. Gonzalez and Woods (1992) described the concept of image segmentation as that segmentation subdivides an image into its constituent parts or objects, and the level to which this subdivision is carried depends on the problem being solved, that is, the segmentation should stop when the objects of interest in an application have been isolated. For example, in beef marbling scoring applications, the interest lies in identifying beef muscles in a beef sample. The first step is to segment the beef sample from the image and then to segment muscle from fat within the sample. The segmentation may stop at the level of separating the muscle from marbling flecks.

Image segmentation techniques are generally based on either discontinuity or similarity of gray-level values. These gray-level discontinuity-based techniques partition an image based on abrupt changes in gray level. Those gray-level similarity-based techniques are based on thresholding, region growing, and region splitting and merging.

Thresholding is one of the effective techniques in image segmentation. The assumption of the thresholding technique in image segmentation is that the object and background pixels in a digital image can be distinguished by their gray-level values. With the optimal threshold, it is possible to divide the image into two gray levels that correspond to the background and the object. Therefore, selection of the threshold value is critical to ensure the successful use of this technique in image segmentation.

There are a number of methods for the determination of the optimal threshold values for image segmentation. Among the methods, image contour lines, local minima in an image histogram, and the magnitude of the gradient of images have been used widely.

Contour lines for an image can be drawn to represent discrete levels of intensity, where each of the lines represents pixels having the same value. The pixels included in the contour line have a value larger than the value represented by the contour line. Then, the image can be segmented into regions by classifying all of the pixels lying below a certain contour line as the object of interest and those having intensities greater than the contour line as the background pixels. This algorithm proceeds in two steps.

First, the contour lines are drawn for the image by dividing the image into a number of equal spaces. Second, based on the contour lines, the optimal threshold for segmenting an interested object is selected, and a binary image is generated.

Gray-level histogram is a simple method for threshold selection. The optimal threshold is determined by optimizing some criterion functions obtained from the gray-level distribution of image. Let $f(x, y)$ be the gray value of the pixel located at the point (x, y) in a digital image $\{f(x, y) \,|\, m \in \{1, 2,..., M\}, n \in \{1, 2,..., N\}\}$ of size $M \times N$, let the histogram be $h(x)$ for $x \in \{0, 1, 2,..., 255\}$, t be a threshold value, and $B = \{b_1, b_0\}$ be a pair of binary gray levels where $b_1, b_0 \in \{0, 1, 2,..., 255\}$. The result of thresholding an image function $f(x, y)$ at gray level T is a binary function $f_T(x, y)$ such that

$$f_T(x, y) = \begin{cases} b_0 & \text{if } f(x, y) > T \\ b_1 & \text{if } f(x, y) \leq T \end{cases} \tag{3.15}$$

Thus, pixels labeled b_0 correspond to objects, whereas pixels labeled b_1 correspond to the background. In general, a thresholding method determines the value T^* of t based on a certain criterion function. If T^* is determined solely from the gray level of each pixel, then the thresholding method is point dependent (Cheng and Tsai, 1993). Over the years, many researchers in image processing have treated $\{f(x, y) \,|\, x \in \{1, 2,..., M\}, n \in \{1, 2,..., N\}\}$ as a sequence of independent, identically distributed random variables with the density function $H(x)$. The density function $H(x)$ can be obtained from

$$H(x) = \text{prob}[f(x, y) = x] \tag{3.16}$$

where $x \in \{0, 1, 2,..., 255\}$. Given an image, the density function can be estimated using the method of relative frequency. In this method, the density function (or the normalized histogram) $H(x)$ is approximated by using the formula

$$h(x) = \frac{N_{pi}(x)}{N_p} \tag{3.17}$$

where $h(x)$ denotes the estimate of $H(x)$, $N_{pi}(x)$ represents the number of pixels with the gray value x, and N_p is the number of pixels in the image. The value of the threshold can be selected from the local minimal, or "valleys," in the histogram.

Usually, the objects contained in an image have a different range of gray levels than the background. If a histogram is plotted for the gray levels, the object and background subparts yield distinct peaks on the histogram. A threshold can usually be selected at the bottom of the valley between these peaks. This algorithm involves computing histograms for gray-level values and then applying thresholds at the local minima or "valleys" (Perwitt and Mendlesohn, 1966). The details can be described as follows

1. Filter the pixels whose values are above the mean value in the image.
2. Generate the histogram.
3. Smooth histogram and take derivations.
4. Apply thresholds at local minima (zero-crossing).

Magnitude of the gradient of an image is another thresholding method based on the gradient of an image, which is similar to edge detection in gray levels. An edge is the boundary between two regions with relatively distinct gray level properties. Basically, the idea underlying most edge-detection techniques is the computation of a local derivation operator. The first derivative is positive for an edge transiting from dark to light while it is negative for an edge transiting from light to dark, and zero in areas of a constant gray level. Hence, the magnitude of the first derivative can be used to define the value of the threshold. Usually, the first derivative at any point in an image is obtained by using the magnitude of the gradient at that point. The gradient of an image $f(x, y)$ at location (x, y) is the vector

$$\nabla f = \begin{bmatrix} G_x \\ G_y \end{bmatrix} = \begin{bmatrix} \dfrac{\partial f}{\partial x} \\ \dfrac{\partial f}{\partial y} \end{bmatrix} \tag{3.18}$$

and the gradient vector points to the direction of maximum rate of change of f at (x, y). The magnitude of this vector, generally referred to simply as the gradient and denoted as

$$\nabla F = \mathrm{mag}(\nabla f)$$

$$= [G_x^2 + G_y^2]^{\frac{1}{2}} \tag{3.19}$$

An edge having a relatively high magnitude must lie at a point of relatively high contrast. Thus, it may be considered important for object detection.

This algorithm selects the segmentation threshold by averaging the intensity of those pixels with high gradients, that is, edge pixels (Katz, 1965). The algorithm followed these procedures

1. Calculate the gradient of the image.
2. Generate the gradient histogram, define the nth percentile, and find the set of pixels with high gradient values.
3. Calculate the gradient threshold by averaging the pixel values in this set.
4. Segment the image with the gradient threshold to generate the binary image.

3.3.1.1   Example: Segmentation of elastograms for detection of hard objects in packaged beef rations

Wang (1998) developed and tested three important algorithms: image contour lines, local minima in an image histogram, and the magnitude of the gradient of images on the modeled elastograms in order to detect hard objects within the packaged beef rations. The three algorithms were tested using a group of different samples. The success rate based upon the number of images with the correct number of hard spots detected, the dependence of detection on the control parameter, and the sensitivity of detection to pixel values were used to evaluate these algorithms.

Figure 3.17 is a typical ultrasonic elastogram of beef sample modeling with a hard object in the center of the image. Figure 3.18 shows the contour lines of this elastogram. The solid line is the selected location of the threshold value. Figure 3.19 shows the segmented binary image created by applying the selected threshold to the original elastogram.

Figure 3.20 shows a histogram of the smoothed elastogram in Figure 3.17. Because the valleys were not easy to find, the first derivative was taken on the histogram by convoluting the histogram using a special operator. The location of the zero crossing, as shown by the marker "*", was used as the location of local minima where the value of the first derivative changed from negative to positive. After the threshold was determined, the elastogram was segmented to a binary image, and the hard spot was clearly noticed (Figure 3.21).

Figure 3.22 shows the gradient histogram of the elastogram and Figure 3.23 shows the segmented image of the elastogram.

3.3.2   Image feature extraction

Once an image has been segmented into areas interested in applications, the areas of segmented pixels need to be described and represented for further processing and analysis. In general, a segmented area can be represented in

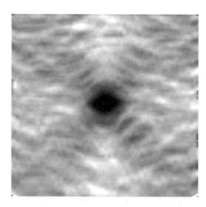

Figure 3.17 A typical ultrasonic elastogram of beef sample modeling containing a hard object in the center of the image. (From Wang, 1998. With permission.)

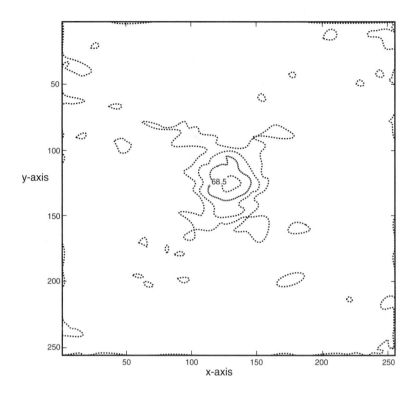

Figure 3.18 Plot of contour lines of the elastogram in Figure 3.17. (From Wang, 1998. With permission.)

terms of its external characteristics like boundary or shape, or its internal characteristics like the values of the pixels in the area. In food quality quantization, information on sample morphology and/or texture from images has been widely used. The extraction of morphological and textual features from images uses the principles of the external and internal representations respectively.

The morphological features of an image are represented as the size and shape of the objects of interest in the image. The term *morphology* originated from a branch of biology that deals with the form and structure of animals and plants. In image processing, the same term is used in the context of mathematical morphology as a method for extracting image components that represent and describe the region shape and the convex hull (Gonzalez and Woods, 1992). In food quality quantization, we are interested in the technique of extraction of morphological (size and shape) features from the images of food samples.

Quantifying the texture content of a segmented area is an important approach to area description. There are three major methods to describe the texture of an area in image processing: statistical, structural, and spectral. Statistical approaches provide textural characteristics such as smoothness, coarseness, graininess, and so on. A typical approach among them was proposed by Haralick et al. (1973). Structural methods deal with the

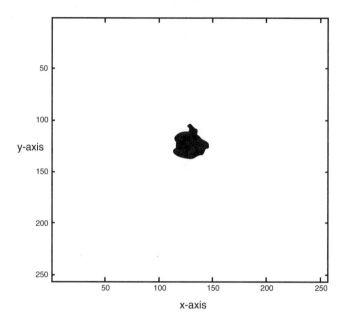

Figure 3.19 The segmented elastogram using the contour line algorithm on the elastogram in Figure 3.17. (From Wang, 1998. With permission.)

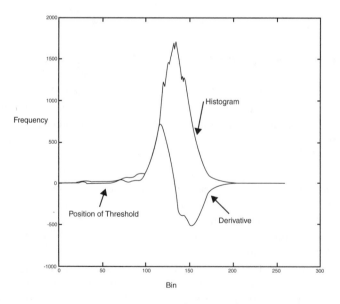

Figure 3.20 Plot of the histogram and the derivative of the histogram of the elastogram in Figure 3.17. (From Wang, 1998. With permission.)

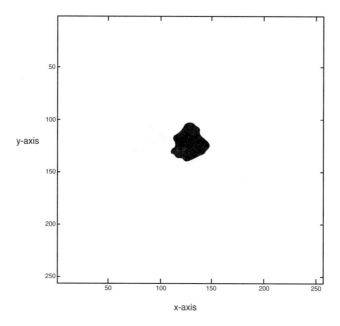

Figure 3.21 The segmented elastogram using the histogram local minima algorithm on the elastogram in Figure 3.17. (From Wang, 1998. With permission.)

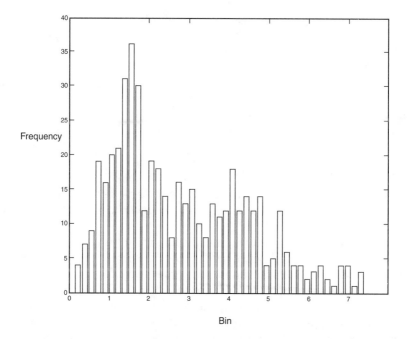

Figure 3.22 Plot of the gradient histogram of the elastogram in Figure 3.17. (From Wang, 1998. With permission.)

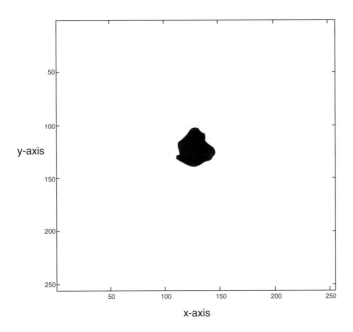

Figure 3.23 The segmented elastogram using the gradient algorithm on the elasto-gram in Figure 3.17. (From Wang, 1998. With permission.)

arrangement of image primitives, such as the description of texture based on regularly spaced parallel lines. Spectral techniques are based on properties of the Fourier spectrum. Wavelet textural analysis (Daubechies, 1988; Mallat, 1989; Huang et al., 1997) can be viewed as a combination of spectral and statistical methods. It is based on the wavelet transform and decomposition of an image for different textural orientations, and then the statistic of each decomposed component is computed as one of the textural features of the image.

Haralick's statistical method (Haralick et al., 1973) consists of 14 easily computable textural features based on graytone spatial dependencies. Haralick et al. (1973) illustrated the theory and applications in category-identification tasks of three different kinds of image data: photomicrographs, aerial pho-tographs, and satellite multispectral imagery. These textural features have been being widely used in the area of food engineering since they were first reported in 1973.

Haralick et al. (1973) assumed that the textural information is adequately represented by the spatial gray-level dependence matrices (co-occurrence matrices) computed for four angular relationship ($\angle\theta = 0°, 45°, 90°$, and $135°$) and at an absolute fixed distance ($d = 1, 2,..., D$) between the neigh-boring resolution cell pairs in an image. Each computed textural feature was derived from the four angular, nearest-neighbor co-occurrence matrices. The co-occurrence matrices are a function of the angular relationship and a func-tion of the absolute distance between neighboring resolution cells and, as a result, these matrices are symmetric.

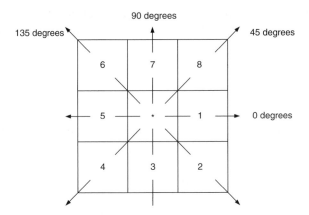

Figure 3.24 Diagram of nearest-neighbor resolution cells. (Adapted from Haralick et al., 1973. With permission.)

The co-occurrence matrices measure the probability that a pixel of a particular gray level will occur at an orientation and a specified distance from its neighboring pixels given that these pixels have a second particular gray level. The co-occurrence matrices can be represented by the function $\mathbf{P}(i, j, d, \angle\theta)$, where i represents the gray level at location (x, y), and j represents the gray level at its neighbor pixel at a distance d and an orientation of $\angle\theta$ from the location (x, y). Figure 3.24 shows eight nearest-neighbor resolution cells, where the surrounding resolution cells are expressed in terms of their spatial orientation to the central reference pixel denoted by the asterisk. In the diagram, the eight neighbors represent all neighbors at a distance of 1. Resolution cells 1 and 5 are the nearest neighbors to the reference cell (*) in the horizontal direction $(\angle\theta = 0°)$ and at a distance of 1 $(d = 1)$. Resolution cells 2 and 6 are the nearest neighbors to the reference cell (*) in the left diagonal direction $(\angle\theta = 135°)$ and at a distance of 1. Resolution cells 3 and 7 are the nearest neighbors to the reference cell (*) in the vertical direction $(\angle\theta = 90°)$ and at a distance of 1. Resolution cells 4 and 8 are the nearest neighbors to the reference cell (*) in the right diagonal direction $(\angle\theta = 45°)$ and at a distance of 1.

Based on the description of the spatial relationship between the resolution cells in the diagram of Figure 3.24, 4 co-occurrence matrices of a 4×4 image $\mathbf{I}(x, y)$ are formed by summing the gray-level values of its neighbor at a specified distance in both directions over the entire image.

Assume

$$\mathbf{I}(x, y) = \begin{bmatrix} 0 & 0 & 1 & 1 \\ 0 & 0 & 1 & 1 \\ 0 & 2 & 2 & 2 \\ 2 & 2 & 3 & 3 \end{bmatrix}$$

This image has four gray tones from 0 to 3.

The general form of the nearest-neighbor resolution cells of a 4×4 image which is used in computing the co-occurrence matrices is as follows

(0, 0)	(0, 1)	(0, 2)	(0, 3)
(1, 0)	(1, 1)	(1, 2)	(1, 3)
(2, 0)	(2, 1)	(2, 2)	(2, 3)
(3, 0)	(3, 1)	(3, 2)	(3, 3)

Then, the four distance 1 angular co-occurrence matrices are computed as

$$
\mathbf{P}(i, j, 1, 0°) = \begin{bmatrix} 4 & 2 & 1 & 0 \\ 2 & 4 & 0 & 0 \\ 1 & 0 & 6 & 1 \\ 0 & 0 & 1 & 2 \end{bmatrix}
$$

$$
\mathbf{P}(i, j, 1, 135°) = \begin{bmatrix} 2 & 1 & 3 & 0 \\ 1 & 2 & 1 & 0 \\ 3 & 1 & 0 & 2 \\ 0 & 0 & 2 & 0 \end{bmatrix}
$$

$$
\mathbf{P}(i, j, 1, 90°) = \begin{bmatrix} 6 & 0 & 2 & 0 \\ 0 & 4 & 2 & 0 \\ 2 & 2 & 2 & 2 \\ 0 & 0 & 2 & 0 \end{bmatrix}
$$

$$
\mathbf{P}(i, j, 1, 45°) = \begin{bmatrix} 4 & 1 & 0 & 0 \\ 1 & 2 & 2 & 0 \\ 0 & 2 & 4 & 1 \\ 0 & 0 & 1 & 0 \end{bmatrix}
$$

Let us look at an example for explaining the computation, the element in the (1, 0) position of the distance 1 horizontal $\mathbf{P}(i, j, 1, 0°)$ matrix is the total number of times 2 gray tones of value 1 and 0 occurred horizontally adjacent to each other. To determine this number, the number of pairs of resolution cells is counted so that the first resolution cell of the pair has gray tone 1, and the second resolution cell of the pair has gray tone 0. Because there are two such pair of (1, 0), the number is 2.

Haralick et al. (1973) presented 14 statistical parameters of image textural features which are computed from each co-occurrence matrix

1. Angular second moment (ASM)

$$
f_1 = \sum_{i=0}^{N_g-1} \sum_{j=0}^{N_g-1} p^2(i, j) \tag{3.20}
$$

where N_g is the number of gray levels from 0 to 255 in the quantified image, $p(i, j) = P(i, j)/R$ for matrix normalization, in which $P(i, j)$ is the co-occurrence matrix, and R is the number of neighboring resolution cell pairs.

ASM is a measure of the homogeneity in an image. A minimal number of values of large magnitude in the co-occurrence matrix is an indication of fewer intensity transitions characteristic of homogeneous images. This usually results in a high ASM, whereas numerous entries of less magnitude in the co-occurrence matrix owing to more intensity transitions usually results in a low ASM.

2. Contrast

$$f_2 = \sum_{d=0}^{N_g-1} d^2 \left(\sum_{i=0}^{N_g-1} \sum_{j=0}^{N_g-1} p(i, j) \right)_{|i-j|=d} \tag{3.21}$$

Contrast is the difference moment of the co-occurrence matrix and is a measure of the amount of local variation presented in an image. A large amount of local variation presented in an image is an indication of high contrast, thereby resulting in higher values for that particular measure.

3. Correlation

$$f_3 = \frac{1}{\sigma_x \sigma_y} \left[\sum_{i=0}^{N_g-1} \sum_{j=0}^{N_g-1} (ij)p(i, j) - \mu_x \mu_y \right] \tag{3.22}$$

where μ_x, μ_y, σ_x, and σ_y are the means and standard deviations of p_x and p_y, while the p_x and p_y are the marginal probability matrices of p.

Correlation is a measure of the linear dependencies of gray-level values (intensities) in an image. Correlation will be much higher for an image with large areas of similar intensities than for an image with noisier, uncorrelated scenes.

4. Variance

$$f_4 = \sum_{i=0}^{N_g-1} \sum_{j=0}^{N_g-1} (i - \mu)^2 \, p(i, j) \tag{3.23}$$

where μ is the mean of p.

Variance is a measure indicating the variation in values of image intensity. Variance would be zero for an image whose pixels all have the same gray level, whereas images containing scenes of variable pixel intensities would result in higher values for variance.

5. Inverse different moment

$$f_5 = \sum_{i=0}^{N_g-1} \sum_{j=0}^{N_g-1} \frac{1}{1+(i-j)^2} p(i, j) \qquad (3.24)$$

Inverse difference moment is another measure of image contrast. The more frequently intensities of similar magnitude occur in an image, the greater the value for this measure.

6. Sum average

$$f_6 = \sum_{i=0}^{2(N_g-1)} (i) p_{x+y}(i) \qquad (3.25)$$

where p_{x+y} is the sum matrix given by

$$p_{x+y}(k) = \left[\sum_{i=0}^{N_g-1} \sum_{j=0}^{N_g-1} p(i, j) \right]_{i+j=k} \qquad (k = 0, 1, 2, ..., 2(N_g - 1))$$

7. Sum variance

$$f_7 = \sum_{i=0}^{2(N_g-1)} (i - f_8)^2 p_{x+y}(i) \qquad (3.26)$$

8. Sum entropy

$$f_8 = -\sum_{i=0}^{2(N_g-1)} p_{x+y}(i) \log[p_{x+y}(i) + \varepsilon] \qquad (3.27)$$

where ε is an arbitrarily small positive constant which is used to avoid $\log(0)$ happening in entropy computation.

Sum entropy is a measure of randomness within an image.

9. Entropy

$$f_9 = \sum_{i=0}^{N_g-1} \sum_{j=0}^{N_g-1} p(i, j) \log[p(i, j) + \varepsilon] \qquad (3.28)$$

Entropy is a measure of the complexity or the amount of orderliness within an image. The more complex (more randomness of gray levels) the image, the higher the entropy value. Low entropy values correspond to high levels of order.

10. Difference variance

$$f_{10} = \text{variance of } p_{x-y} \tag{3.29}$$

where p_{x-y} is the difference matrix given by

$$p_{x-y}(k) = \left[\sum_{i=0}^{N_g-1} \sum_{j=0}^{N_g-1} p(i,j) \right]_{i-j=k} \quad (k = 0, 1, 2, \ldots, N_g - 1).$$

11. Difference entropy

$$f_{11} = - \sum_{i=0}^{N_g-1} p_{x-y}(i) \log[p_{x-y}(i) + \varepsilon] \tag{3.30}$$

Difference entropy is also a measure of the amount of order in image.

12. Informational measure of correlation – 1

$$f_{12} = \frac{HXY - HXY1}{\max[HX, HY]} \tag{3.31}$$

13. Informational measure of correlation – 2

$$f_{13} = \sqrt{1 - \exp[-2.0(HXY2 - HXY)]} \tag{3.32}$$

where HXY is the entropy of p, HX and HY are entropies of p_x and p_x, and

$$HXY1 = - \sum_{i=0}^{N_g-1} \sum_{j=0}^{N_g-1} p(i,j) \log[p_x(i)p_y(j) + \varepsilon]$$

$$HXY2 = - \sum_{i=0}^{N_g-1} \sum_{j=0}^{N_g-1} p_x(i)p_y(j) \log[p_x(i)p_y(j) + \varepsilon]$$

14. Maximal correlation coefficient

$$f_{14} = (\text{second largest eigenvalue of } Q)^{1/2} \tag{3.33}$$

where $q(i,j) = \sum_{k=0}^{N_g-1} [p(i,k)p(j,k)/p_x(i)p_y(k)]$.

Besides the 14 statistical textural feature parameters, other gray-level image features are also extracted and used for image analysis. Whittaker et al. (1992) grouped statistics of image intensity, Fourier transform, fractal dimension, and slope of attenuation extracted from ultrasonic images for prediction of a beef marbling score. Thane (1992) gave a description of five gray-level image intensity features

1. Image intensity measurements—there are three parameters related to measures of image intensity.
    a. Mean of image intensity (*IM*)—intensity mean, as defined by Gonzalez and Wintz (1987), is a measure of the average brightness in an image. The following equation represents a mathematical expression for *IM*

$$\bar{x} = \frac{1}{n} \left[\sum_{i=0}^{N_r-1} \sum_{j=0}^{N_c-1} I(i, j) \right] \tag{3.34}$$

    where N_r is the total number of rows in the image, N_c is the total number of columns in the image, and n is the total number of pixels having intensity levels greater than zero. $\bar{x}$ is the summation of image intensity values ($I(i, j) > 0$, for each pixel.
    b. Sample standard deviation of image intensity (*ISD*)—variance or standard deviation is a measure of image contrast related to the variability of image intensity values about the mean. The expression of the *ISD* is as follows

$$SD = \sqrt{\frac{1}{n-1} \left[\sum_{i=0}^{N_r-1} \sum_{j=0}^{N_c-1} (I(i, j) - \bar{x})^2 \right]} \tag{3.35}$$

    c. Total pixel count (*IC*)—the total count of all pixels in the image having an intensity level greater than a given threshold value ($T = 0$) is computed by the following equation

$$IC = \sum_{i=0}^{N_r-1} \sum_{j=0}^{N_c-1} I(i, j) \tag{3.36}$$

2. Ratio of Euclidean to linear distance of image intensity (*RELI*)—the *RELI* measure can be considered as an indicator of image texture. The computation of *RELI* is performed as follows. First, the horizontal curvilinear distance for each row in the image is determined by adding the linear distances between each pixel intensity in the row and dividing by the number of columns in the image

$$D(i) = \sum_{i=0}^{N_r-1} \left[\frac{1}{N_c} \left(\sum_{j=1}^{N_c-1} \sqrt{|I(i, j) - I(i, j-1)|^2 + 1.0} \right) \right]$$

So, the *RELI* can be calculated by dividing the sum of the values for $D(i)$ by the total number of rows in the image

$$RELI = \frac{1}{N_r} \sum_{i=0}^{N_r-1} D(i) \tag{3.37}$$

RELI is the calculated average horizontal ratio of Euclidean to linear length of the intensity for the image.

3. A measure of slope of attenuation (*SLP*)—the slope of attenuation is one of the most commonly used parameters for estimating ultrasonic attenuation and is calculated through Least Squares regression analysis (Wilson et al., 1984). Milton and Arnold (1986) gave the computing expression of the *SLP*

$$SLP = \frac{N_c \sum_{i=1}^{N_c} x_i y_i - \sum_{i=1}^{N_c} x_i \sum_{i=1}^{N_c} y_i}{N_c \sum_{i=1}^{N_c} x_i^2 - \left(\sum_{i=1}^{N_c} x_i\right)^2} \tag{3.38}$$

where y is the average value of intensity for each corresponding value of x in the image.

Each of these features is primarily image-content dependent and was computed for an 80×80 AOI (area of interest) extracted from the original ultrasonic images.

For the 14 statistical textural features, 4 angular ($\angle\theta = 0°$, $45°$, $90°$, and $135°$) co-occurrence matrices were derived for each of the 8 nearest neighbors for each image. Distinct distances of $d = 1$, 2, and 3 were considered.

Wavelet decomposition is another way for image textural feature extraction. Wavelet bases that are two-dimensional can be constructed from one-dimensional bases by the tensor product (Mallat, 1989). For the scaling function, it is

$$\Phi(x, y) = \Phi(x)\Phi(y) \tag{3.39}$$

For wavelets, they are

$$\Psi_1(x, y) = \Phi(x)\Psi(y) \tag{3.40}$$

$$\Psi_2(x, y) = \Psi(x)\Phi(y) \tag{3.41}$$

$$\Psi_3(x, y) = \Psi(x)\Psi(y) \tag{3.42}$$

For a two-dimensional signal $f(x, y)$, the resulting equations of the extension of the Mallat's algorithm to the two-dimensional wavelet transform are

$$f^{(m)}(k_x, k_y) = \sum_{i_y}\sum_{i_x} f^{(m+1)}(i_x, i_y)h(2k_x - i_x)h(2k_y - i_y) \qquad (3.43)$$

$$d_1^{(m)}(k_x, k_y) = \sum_{i_y}\sum_{i_x} f^{(m+1)}(i_x, i_y)g(2k_x - i_x)h(2k_y - i_y) \qquad (3.44)$$

$$d_2^{(m)}(k_x, k_y) = \sum_{i_y}\sum_{i_x} f^{(m+1)}(i_x, i_y)h(2k_x - i_x)g(2k_y - i_y) \qquad (3.45)$$

$$d_3^{(m)}(k_x, k_y) = \sum_{i_y}\sum_{i_x} f^{(m+1)}(i_x, i_y)g(2k_x - i_x)g(2k_y - i_y) \qquad (3.46)$$

where k_x and i_x are sample points on the x axis, and k_y and i_y are sample points on the y axis. In the case of two dimensions, a smoothed signal $f^{(m)}$ and three detail signals, $d_1^{(m)}$, $d_2^{(m)}$, and $d_3^{(m)}$, which represent the detail signal array in the x, y, and diagonal directions, respectively, are obtained.

The two-dimensional wavelet transform algorithm decomposes a two-dimensional signal array into smoothed and detail signals at different resolutions. For extraction of image textural features, this algorithm can be applied directly to images of food samples. The implementation of this algorithm is row convolutions followed by column subsamplings at one-half rate, then column convolutions followed by row subsamplings at one-half rate. The subsamplings are in increasing scales until the image is reduced to a single pixel.

When using the two-dimensional wavelet decomposition algorithm to extract textural features from each image, the number of features is dependent on the size of the image. If the size of an image is $N \times N$ ($N = 2^M$), the wavelet decomposition of the image has M levels (resolutions) and $4 \times M + 1$ blocks for feature generation. Figure 3.25 shows that the original image is set as the initial smoothed component and then, at each level, the smoothed component is decomposed into four subcomponents: smoothed, vertical, horizontal, and diagonal, at a coarser resolution until the smoothed component is a single pixel. In performing the wavelet decomposition, at each level the smoothed component is transformed in rows first and then in columns. This is shown in Figure 3.26. It can be shown that $H_cH_rE^{(m)}$ is the smoothed component at the next lower resolution, and three other components, $G_cG_rE^{(m)}$, $H_cG_rE^{(m)}$, and $G_cH_rE^{(m)}$ are details.

Color information is also useful in quantifying the images of food samples. The attributes of color information in images are R (red), G (green), and B (blue) or H (hue), S (saturation), and I (intensity). This book does not intend to detail color image processing. Interested readers can refer to the

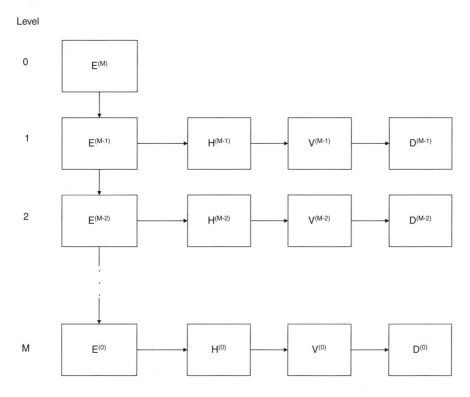

Figure 3.25 Wavelet decomposition scheme for image feature extraction, where $E^{(M)}$ represents the original image, and $E^{(j)}$, $H^{(j)}$, $V^{(j)}$, and $D^{(j)}$ ($j = M - 1,..., 1, 0$) are smoothed, horizontal, vertical, and diagonal components, respectively, at each level. (Adapted from Huang et al., 1997. With permission.)

book by Gonzalez and Woods (1992) for theoretical fundamentals, and the paper by Gerrard et al. (1996) for the application in food quality quantization.

3.3.2.1 Example: Morphological and Haralick's statistical textural feature extraction from images of snack food samples

In the study of the evaluation of the quality of typical snack food products (Sayeed et al., 1995), the size and shape features together with the external texture features that can reflect the internal structure of the snack images were used to describe the quality from a texture (mouthfeel) standpoint.

The machine vision system described in the last chapter captured the images of the snacks and chips. Then, in order to quantify the quality of the snack products in terms of morphology and texture, certain operations were performed on the images.

To obtain the morphological features, the images were thresholded based on their histograms. The binary images were then formed and processed by a closing morphology operation with a disk structuring element to obtain the

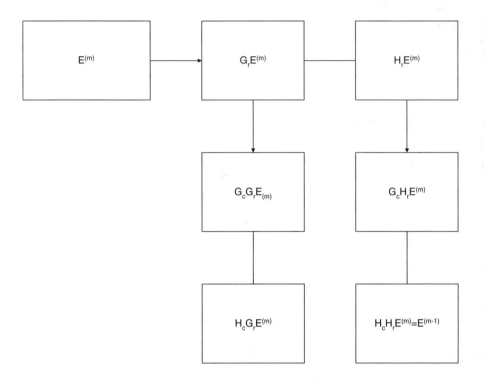

Figure 3.26 Transform scheme in wavelet decomposition for image feature extraction, where H_r and G_r are the row transforms associated with the $h(i)$ and $g(i)$ filters, respectively, and H_c and G_c are the column transforms associated with the $h(i)$ and $g(i)$ filters, respectively. (Adapted from Huang et al., 1997. With permission.)

size and shape features (Gonzalez and Woods, 1992). The following nine measurable features were used to describe the size and shape of the snacks:

1. Area (AREA)—the number of pixels contained in a snack object; this number is converted into a physical size by using the calibration parameter of the camera.
2. Perimeter (PERIM)—the number of pixels along the boundary of a snack object; again, the calibration parameter is used to compute the corresponding physical length.
3. Fiber length (FIBERL) and width (BREADTH)—fiber length and width are considered to be the length and width of a rectangle surrounding a snack object.
4. Length (LENGTH) and breadth (BREADTH)—length is considered to be the longest chord passing through a snack object. Breadth is the shortest chord passing through it.
5. Roundness (ROUND)—this is a shape factor which has a minimum value of 1.0 for a circular shape snack. Large values of roundness indicate thinner and longer snacks.

6. Fullness ratio (FULLR)—fullness ratio is the ratio of the snack area to the circumscribed area.
7. Aspect ration (ASPR)—this is the ratio of the length to the breadth of a snack object.

Except for the maximal correlation coefficient, 13 of the 14 textural features as defined by Haralick et al. (1973) were calculated based on co-occurrence matrices reflecting the spatial distribution of intensity variations from the images of the snacks:

1. Angular second moment (F_1).
2. Contrast (F_2).
3. Correlation (F_3).
4. Variance (F_4).
5. Inverse difference moment (F_5).
6. Sum average (F_6).
7. Sum variance (F_7).
8. Sum entropy (F_8).
9. Entropy (F_9).
10. Difference variance (F_{10}).
11. Difference entropy (F_{11}).
12. Information measure of correlation number 1 (F_{12}).
13. Information measure of correlation number 2 (F_{13}).

A total of 22 features, 9 morphological and 13 textural, from snack images were extracted as the quality evaluation process input parameters to correlate with 7 sensory attributes as the process output parameters that define the visual quality of the snack products by the taste panel

1. Bubble.
2. Roughness.
3. Cell size.
4. Firmness.
5. Crispiness.
6. Tooth packing.
7. Grittiness of mass.

The taste panel scaled these sensory attributes in the range of −3 to 3 with 0 indicating the optimum value.

3.3.2.2 *Example: Feature extraction from ultrasonic B-mode images for beef grading*

Besides 14 statistical textural feature parameters, 5 gray-level image intensity features were also performed on the AOI of ultrasonic images for beef marbling prediction purposes (Thane, 1992). There were a total of 19 features extracted from the beef ultrasonic images. The five gray-level image intensity features included image intensity measurements *IM*, *ISD*, and *IC*, *RELI*, and *SLP*. Each of these features is primarily image-content dependent and was computed for an 80 × 80 AOI extracted from the original ultrasonic images.

For the 14 statistical textural features, 4 angular ($\angle\theta = 0°$, 45°, 90°, and 135°) co-occurrence matrices were derived for each of the 8 nearest neighbors for each image. Distinct distances of $d = 1$, 2, and 3 were considered.

3.3.2.3   Example: Haralick's statistical textural feature extraction from meat elastograms

For meat quality evaluation, elastograms of beef and pork samples were used to produce parameters of textural features using the method of Haralick et al. (1973) (Lozano, 1995; Moore, 1996). In the extraction of textural features, for each elastographic image, all 14 parameters originally presented by Haralick et al. (1973) were computed for four angles (0°, 45°, 90°, and 135°) and four neighborhood distances ($d = 1$, 2, 5, and 10). There are 16 (4 × 4) independent groups of statistical textural features, each of which contains the 14 parameters, for each meat elastogram. Figure 3.27 shows a typical output list from the computation of Haralick's statistical textural features of a beef elastogram.

The 14 parameters were used as the input of the quality evaluation process to correlate with 10 mechanical and chemical variables as the output of the quality evaluation process

1. Warner-Bratzler shear force at 2 days (WB1).
2. Warner-Bratzler shear force at 14 days (WB2).
3. Warner-Bratzler shear force at 28 days (WB3).
4. Warner-Bratzler shear force at 42 days (WB4).
5. Calpastatin (Calp).
6. Sarcomere length (Sarc).
7. Total collagen amount (T.Coll).
8. Soluble collagen in percentage (%Sol).
9. Percent moisture (%Mois).
10. Percent fat (%Fat).

3.3.2.4   Example: Wavelet textural feature extraction from meat elastograms

For meat quality prediction based on the technique of elastography, Huang et al. (1997) developed the method of wavelet decomposition for extraction of textural features from elastograms. Figure 2.15 in the last chapter shows a beef elastogram of LD muscle. This image is used to explain the process of wavelet decomposition for this application. First, the image was resized to the power of 2 in order to meet the requirement of wavelet analysis. Figure 3.28 is the resized elastogram. The size of the elastogram now is 128 × 128 compared to the original size 134 × 198.

distance = 1

Angle	0	45	90	135	Avg
Angular Second Moment	1.319e-03	1.014e-03	1.535e-03	1.007e-03	1.219e-03
Contrast	4.254e+01	6.288e+01	2.492e+01	6.349e+01	4.846e+01
Correlation	2.411e+08	2.389e+08	2.473e+08	2.388e+08	2.415e+08
Variance	9.770e+02	9.780e+02	9.871e+02	9.779e+02	9.800e+02
Inverse Diff Moment	2.383e-01	1.788e-01	2.594e-01	1.784e-01	2.138e-01
Sum Average	5.617e+01	5.621e+01	5.637e+01	5.621e+01	5.624e+01
Sum Variance	3.642e+03	3.626e+03	3.699e+03	3.625e+03	3.648e+03
Sum Entropy	2.023e+00	2.019e+00	2.027e+00	2.018e+00	2.022e+00
Entropy	3.029e+00	3.132e+00	2.968e+00	3.133e+00	3.066e+00
Difference Variance	2.006e-05	1.557e-05	2.216e-05	1.547e-05	1.831e-05
Difference Entropy	1.083e+00	1.191e+00	1.031e+00	1.192e+00	1.125e+00
Means of Correlation-1	-.2521e-01	-1.925e-01	-2.898e-01	-1.918e-01	-2.316e-01
Means of Correlation-2	7.633e-01	6.978e-01	7.964e-01	6.968e-01	7.386e-01
Max Correlation Coeff	1.650e-03	4.050e-03	1.180e-02	1.937e-02	9.219e-03

distance = 2

Angle	0	45	90	135	Avg
Angular Second Moment	8.618e-04	6.676e-04	8.681e-04	6.638e-04	7.653e-04
Contrast	9.309e+01	1.389e+02	8.105e+01	1.400e+02	1.133e+02
Correlation	2.348e+08	2.305e+08	2.414e+08	2.303e+08	2.342e+08
Variance	9.784e+02	9.832e+02	9.908e+02	9.832e+02	9.839e+02
Inverse Diff Moment	1.657e-01	1.176e-01	1.560e-01	1.185e-01	1.394e-01
Sum Average	5.624e+01	5.641e+01	5.650e+01	5.641e+01	5.639e+01
Sum Variance	3.599e+03	3.573e+03	3.659e+03	3.572e+03	3.600e+03
Sum Entropy	2.008e+00	1.995e+00	2.014e+00	1.995e+00	2.003e+00
Entropy	3.214e+00	3.300e+00	3.205e+00	3.304e+00	3.256e+00
Difference Variance	1.306e-05	9.889e-06	1.245e-05	9.913e-06	1.133e-05

	0	45	90	135	Avg
Difference Entropy	1.282e+00	1.384e+00	1.279e+00	1.385e+00	1.333e+00
Means of Correlation-1	-1.445e-01	-9.507e-02	-1.535e-01	-9.309e-02	-1.215e-01
Means of Correlation-2	6.276e-01	5.298e-01	6.427e-01	5.251e-01	5.813e-01
Max Correlation Coeff	6.503e-03	7.247e-03	2.519e-02	1.363e-02	1.314e-02

distance = 5

Angle	0	45	90	135	Avg
Angular Second Moment	6.122e-04	5.086e-04	5.230e-04	5.165e-04	5.401e-04
Contrast	1.664e+02	2.411e+02	2.621e+02	2.426e+02	2.281e+02
Correlation	2.270e+08	2.234e+08	2.229e+08	2.238e+08	2.243e+08
Variance	9.837e+02	1.005e+03	1.007e+03	1.006e+03	1.000e+03
Inverse Diff Moment	1.097e-01	8.079e-02	8.668e-02	8.697e-02	9.105e-02
Sum Average	5.643e+1	5.720e+01	5.709e+01	5.723e+01	5.699e+01
Sum Variance	3.548e+03	3.557e+03	3.547e+03	3.562e+03	3.554e+03
Sum Entropy	1.985e+00	1.959e+00	1.948e+00	1.955e+00	1.962e+00
Entropy	3.335e+00	3.397e+00	3.396e+00	3.395e+00	3.381e+00
Difference Variance	8.975e-06	7.103e-06	6.754e-06	7.143e-06	7.494e-06
Difference Entropy	1.426e+00	1.511e+00	1.531e+00	1.512e+00	1.495e+00
Means of Correlation-1	-7.463e-02	-3.925e-02	-4.410e-02	-4.134e-02	-4.983e-02
Means of Correlation-2	4.773e-01	3.566e-01	3.768e-01	3.654e-01	3.940e-01
Max Correlation Coeff	4.565e-03	1.815e-02	1.506e-03	1.235e-02	9.145e-03

distance = 10

Angle	0	45	90	135	Avg
Angular Second Moment	5.630e-04	4.703e-04	4.932e-04	4.966e-04	5.058e-04
Contrast	2.025e+02	3.037e+02	3.248e+02	2.813e+02	2.781e+02
Correlation	2.244e+08	2.259e+08	2.234e+08	2.299e+08	2.259e+08
Variance	9.918e+02	1.043e+03	1.036e+03	1.048e+03	1.030e+03
Inverse Diff Moment	9.902e-02	6.783e-02	7.636e-02	7.733e-02	8.013e-02
Sum Average	5.672e+01	5.855e+01	5.813e+01	5.871e+01	5.803e+01

Sum Variance	3.545e+03	3.645e+03	3.601e+03	3.68e+03	3.619e+03
Sum Entropy	1.971e+00	1.930e+00	1.919e+00	1.942e+00	1.941e+00
Entropy	3.368e+00	3.418e+00	3.415e+00	3.407e+00	3.402e+00
Difference Variance	7.928e-06	6.229e-06	6.049e-06	6.671e-06	6.719e-06
Difference Entropy	1.472e+00	1.561e+00	1.577e+00	1.543e+00	1.538e+00
Means of Correlation-1	-5.524e-02	-2.771e-02	-3.433e-02	-3.419e-02	-3.787e-02
Means of Correlation-2	4.173e-01	3.026e-01	3.353e-01	3.343e-01	3.474e-01
Max Correlation Coeff	7.902e-03	1.264e-02	7.117e-03	8.468e-03	9.031e-03

Figure 3.27 An output list of Haralick's statistical textural features of a beef elastogram from a C language program.

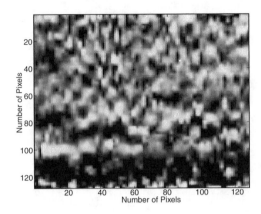

Figure 3.28 The resized elastogram of beef in LD muscle. (From Huang et al., 1997. With permission.)

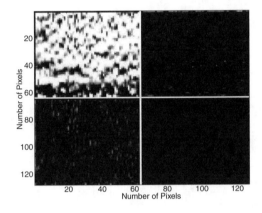

Figure 3.29 The first-step, second-order Daubechies wavelet-transformed beef elastogram of the image in Figure 3.28. (From Huang et al., 1997. With permission.)

In the process of decomposition, the second order Daubechies wavelet was used. As shown in Figure 3.29, at the first step of the decomposition process, the upper left quarter in the transformed image approximates the image of Figure 3.28 by a downsampling at 2; the upper right quarter in the transformed image captures the horizontal textures of the image of Figure 3.28 by a downsampling at 2; the lower left quarter in the transformed image captures the vertical textures of the image of Figure 3.28 by a downsampling at 2; and the lower right quarter in the transformed image captures the diagonal textures of the image of Figure 3.28 by a downsampling at 2. In this way, the first-step second-order Daubechies wavelet-transformed elastogram was built up in Figure 3.29. In this and later images, the bright straight lines

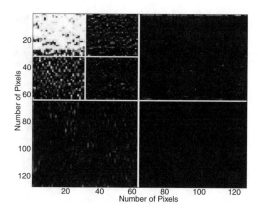

Figure 3.30 The second-step, second-order Daubechies wavelet-transformed beef elastogram of the image in Figure 3.28. (From Huang et al., 1997. With permission.)

are used to describe the decomposition frame, from which the relationships between the four quarters and the image of Figure 3.28 can be seen clearly.

At the second step, the decomposition components in the upper right, lower left, and lower right quarters in the transformed image of Figure 3.29 remain unchanged because they are all detail components. The decomposition is enacted only on the smoothed part (upper left quarter) of the transformed image of Figure 3.29. As shown in Figure 3.30, the upper left quarter of the upper left quarter in the transformed image approximates the smoothed part (upper left quarter) of Figure 3.30 by a downsampling at 2; the upper right quarter of the upper left quarter in the transformed image captures the horizontal textures of the smoothed part (upper left quarter) of Figure 3.30 by a downsampling at 2; the lower left quarter of the upper left quarter in the transformed image captures the vertical textures of the smoothed part (upper left quarter) of Figure 3.29 by a downsampling at 2; and the lower right quarter of the upper left quarter in the transformed image captures the diagonal textures of the smoothed part (upper left quarter) of Figure 3.29 by a downsampling at 2. In this way, the second-step second-order Daubechies wavelet-transformed elastogram was further built up in Figure 3.30. With the help of the bright straight lines for the decomposition frame, the relationships between the transformed images of Figures 3.29 and 3.30 can be seen clearly.

We continued this decomposition until the lowest resolution (a single pixel) was reached. Figure 3.31 shows the complete wavelet-transformed elastogram with the second-order Daubechies basis. In the transformed elastogram, all detail components and one smoothed component are in the single pixel at the left top corner. In this transformed elastogram, a series of the upper right quarters were reduced in size by $\frac{1}{4}$ until the last pixel at the upper left corner was reached. These quarters captured the horizontal textures. Likewise, a series of the lower left quarters were reduced in size by $\frac{1}{4}$ until the last

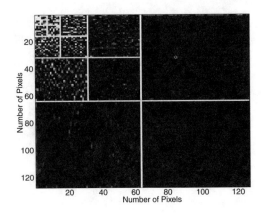

Figure 3.31 The complete second-order Daubechies wavelet-transformed beef elastogram of the image in Figure 3.28. (From Huang et al., 1997. With permission.)

pixel at the upper left corner was reached. These quarters captured the vertical textures. A series of the lower right quarters was reduced in size by $\frac{1}{4}$ until the last pixel at the upper left corner was reached. These quarters captured the diagonal textures.

In the decomposition of the second-order Daubechies wavelets, boundary problems occurred because the image signals were only specified over a finite interval. In order to handle the problem, the symmetric boundary conditions (Aldroubi and Unser, 1996) were used (Huang et al., 1997). Because of the boundary problem, handling the quarters in image decomposition were extended. Consequently, in the preceding image displays for wavelet decomposition, some extended pixels were removed to fit the quarter size reduction in the power of 2.

In this way, a given elastogram was decomposed into smoothed and detailed quarter components corresponding to the blocks as described in Figure 3.25. Each of these components or blocks can be used to generate a texture feature parameter. In this application, the feature parameter was defined as the root mean square (*RMS*)

$$RMS = \sqrt{\sum_{ij} PV_{ij}^2} \qquad (3.47)$$

where PV_{ij} is the pixel value in a quarter component, and all squared PVs in the quarter component are summed. With this parameter, for a 128×128 elastogram, there are 29 ($= 4 \times 7 + 1$) RMS wavelet textural feature parameters.

The 29 parameters were used as the input of the quality evaluation to correlate with 10 mechanical and chemical parameters as the output of the quality evaluation process described previously.

References

Aldroubi, A. and Unser, M., *Wavelets in Medicine and Biology*, CRC Press, Boca Raton, FL, 1996.

Cheng, S. C. and Tsai, W. H., A neural network implementation of the moment-preserving technique and its applications to thresholding, *IEEE Trans. Comput.*, 42, 501, 1993.

Daubechies, I., Orthonormal bases of compactly supported wavelets, *Comm. Pure Appl. Math.*, 41(7), 909, 1988.

Gerrard, D. E., Gao, X., and Tan, J., Beef marbling and color score determination by image processing, *J. Food Sci.*, 61(1), 145, 1996.

Gonzalez, R. C. and Woods, R. E., *Digital Image Processing.*, Addison-Wesley Reading, MA, 1992.

Gonzalez, R. C. and Wintz, P., *Digital Image Processing*, 2nd ed., Addison-Wesley, Reading, MA, 1987.

Haralick, R. M., Shanmugam, K., and Dinstein, I., Textural features for image classification, *IEEE Trans. Syst., Man, Cybern.*, 3(6), 610, 1973.

Huang, Y., Snack food frying process input–output modeling and control through artificial neural networks, Ph.D. dissertation, Texas A&M University, College Station, TX, 1995.

Huang, Y. and Whittaker, A. D., Input–output modeling of biological product processing through neural networks, ASAE paper No. 93-3507. St. Joseph, MI, 1993.

Huang, Y., Lacey, R. E., Moore, L. L., Miller, R. K., Whittaker, A. D., and Ophir, J., Wavelet textural features from ultrasonic elastograms for meat quality prediction, *Trans. ASAE*, 40(6), 1741, 1997.

Katz, Y. H., Pattern recognition of meteorological satellite photography, *Proc. Symp. Remote Sens. Environ.*, 12, 173, 1965.

Lacey, R. E. and Osborn, G. S., Applications of electronic noses in measuring biological systems, ASAE paper No. 98-6116, St. Joseph, MI, 1998.

Lozano, M. S. R., Ultrasonic elastography to evaluate beef and pork quality, Ph.D. dissertation, Texas A&M University, College Station, TX, 1995.

Mallat, S., A theory of multiresolution signal decomposition: the wavelet representation, *IEEE Trans. Patt. Anal. Mach. Intell.*, 11(7), 674, 1989.

Milton, J. S. and Arnold, J. C., *Probability and Statistics in the Engineering and Computing Sciences*, McGraw-Hill, New York, 1986.

Moore, L. L., Ultrasonic elastography to predict beef tenderness, M.Sc. thesis, Texas A&M University, College Station, TX, 1996.

Osborn, G. S., Lacey, R. E., and Singleton, J. A., A method to detect peanut off-flavors using an electronic nose, *Trans. ASAE*, in press.

Osborn, G. S., Lacey, R. E., and Singleton, J. A., Detecting high temperature curing off-flavors in peanuts using an electronic nose, ASAE Paper no. 986075, 1998.

Park, B., Non-invasive, objective measurements of intramuscular fat in beef through ultrasonic A-model and frequency analysis, Ph.D. dissertation, Texas A&M University, College Station, TX, 1991.

Perwitt, J. and Mendlesohn, M., The analysis of cell images, *Ann. N.Y. Acad. Sci.*, 135, 1035, 1966.

Sayeed, M. S., Whittaker, A. D., and Kehtarnavaz, N. D., Snack quality evaluation method based on image features and neural network prediction, *Trans. ASAE*, 38(4), 1239, 1995.

Thane, B. R., Prediction of intramuscular fat in live and slaughtered beef animals through processing of ultrasonic images, M.Sc. thesis, Texas A&M University, College Station, TX, 1992.

Wang, W., Development of elastography as a non-invasive method for hard spots detection of packaged beef rations, M.Sc. thesis, Texas A&M University, College Station, TX, 1998.

Whittaker, A. D., Park, B., Thane, B. R., Miller, R. K., and Savell, J. W., Principles of ultrasound and measurement of intramuscular fat, *J. Anim. Sci.*, 70, 942, 1992.

Wilson, L. S., Robinson, D. E., and Doust, B. D., Frequency domain processing for ultrasonic attenuation measurement in liver, *Ultrason. Imag.*, 6, 278, 1984.

chapter four

Modeling

4.1 Modeling strategy

Data analysis reveals qualitative, noncausal relationships between variables. For example, data analysis can determine that a variable x is closely related to a variable y, but it cannot express how a variable affects another variable. If classification or prediction of a variable is needed, then the causal relationship between this variable and affecting variable(s) must be characterized. In other words, one must set up a quantitative relationship to characterize how one variable affects the other. This is the goal of modeling. In modeling, the affected variable is called a dependent variable, usually an output variable, and the affecting variable is called an independent variable, usually an input variable. In food quality quantization, based on data analysis, modeling is performed to establish quantitative relationships between inputs and outputs using mathematical and statistical methods. Modeling is critical for effective food quality classification, prediction, and process control.

The meaning of process is broad. It can be a manufacturing system, a biological system, and an economic or sociological system as well. The processes in this book are in food manufacturing and processing; for example, processes of extrusion and frying in the snack food industry. This chapter focuses on how to build mathematical models of a process in order to better analyze process mechanism, predict and understand process performance, and design effective process control systems for food quality assurance.

4.1.1 Theoretical and empirical modeling

A model is a description of the useful information extracted from a given practical process. This description is about the mechanism of process operation and it is an abstract or simplification of the practice. Modeling is a tool for data analysis, classification, prediction, and process control. In general, there are two methods in modeling: theoretical and empirical.

Theoretical modeling is the process of building a mathematical model about a relationship through analyzing the operating mechanism and using known laws, theorems, and principles, such as chemical kinetic principles, biological laws, Newtonian laws, material equilibrium equations, energy equilibrium equations, thermodynamic equations, heat and mass transfer principles, and so on. Theoretical modeling is often used in food analysis. For example, Moreira et al. (1995) modeled deep-fat frying of tortilla chips with a mathematical description of heat transfer coupled to the transport of mass. However, this method is limited for complex relationships and processes in the food industry because theoretical modeling is based on assumptions of reasonable simplification of the concerned relationship or process or the problem may get to be too complicated. Also, it is a difficult task to make these assumptions fit the practical situation, and some mechanism of practical process may be unknown. In addition, some factors in a process may change constantly, making it difficult to describe it precisely. For such reasons, a different modeling method is needed to handle the uncertainty and the unknowns.

Empirical modeling has been widely used in process modeling and control. Empirical modeling is used when the relationship between variables is difficult to describe or implement by theoretical methods. In this case, the details about the links between variables usually are set aside, and the focus is put on the effect of input variables to output variables. A typical procedure for empirical modeling is as follows:

1. Model hypothesis—in this step, a model structure for system input and output is hypothesized. A typical one is the simple regression model

$$y = ax + b + \varepsilon \qquad (4.1)$$

   where y is the system output, x is the system input, a and b are model coefficients, and ε is the term of modeling residuals, representing other affecting factors that are minor compared to x, and ε is also assumed to be an independent random variable with zero mean and the same variance as y.

   In this hypothesized model, one system input, x, is assumed to linearly relate to one system output, y, with a slope, a, and an intercept, b, in a straight line.

2. Model estimation—in this step, the values of the slope, a, and the intercept, b, are chosen to make the term residuals, ε, minimal. This can be done by the method of least squares (Milton and Arnold, 1990) with the experimental data of x and y to get the estimated values of $\hat{a}$ and $\hat{b}$.

3. Model test—in this step, the significance of the estimated values of $\hat{a}$ and $\hat{b}$ is tested. This may be done by a t test when the sample size is less than 30 or by Z test when the sample size is more than 30 (Milton and Arnold, 1990). If the test passes, that is, $\hat{a}$ and $\hat{b}$ are significantly different from 0, go to the next step; otherwise go back to Step 1 to adjust the model hypothesis.

4. Model application—in this step, the model is built up for classification, prediction, or process control. The model is

$$y = \hat{a}x + \hat{b} \tag{4.2}$$

The preceding model is a simple one for the case of SISO. Actually, it is not difficult to extend it to the case of MIMO as long as y, x, a, b, and ε are set as vectors $\underline{y}$, $\underline{x}$, $\underline{a}$, $\underline{b}$, and $\underline{\varepsilon}$. Also, it is not necessary for the relationship between inputs and outputs to be linear. It could be in some form of non-linearity to capture the causality.

4.1.2 Static and dynamic modeling

Static modeling is often performed for food quality classification and attribute prediction. In many cases of food quality modeling, the relationship between system inputs and outputs does not take the dimension of time into account, which means that the constant relationship is sought and the inputs and outputs are statically related. The systems of this kind are generally assumed to be governed by the equation

$$\underline{y} = f(\underline{x}, \Theta) + \underline{\varepsilon} \tag{4.3}$$

where $\underline{y} = (y_1, y_2,...,y_m)^T$ is an m-dimension vector representing m system outputs in the model assumption, $\underline{x} = (x_1, x_2,..., x_n)^T$ is an n-dimension vector representing n system inputs in the model assumption, $\underline{\varepsilon} = (\varepsilon_1, \varepsilon_2,..., \varepsilon_m)^T$ is an m-dimension vector representing m system residual variables corresponding to system outputs, Θ represents the set of coefficients in the model (in the linear case, Θ should be a vector of size $n + 1$ while in the nonlinear case, it depends on the model structure and $f(\ )$ is the function describing how the inputs and outputs relate.

Dynamic modeling is necessary in process modeling for process control. In many other cases of food quality modeling, especially those for food process control, the models need to take into account, the dimension of time that is, the relationship changes with time and the system inputs and outputs are related through time. Thus, the variables that are connected in this way relate dynamically to each other. Dynamic models are used to describe the relationships between state variables in a transient stage. The models need to characterize the changes in outputs caused by the changes in inputs to make the system move from one state to the other. The dynamic systems generally can be assumed as following the equation

$$\underline{y}(t) = f(\underline{y}(t-1), \underline{y}(t-2),..., \underline{y}(t-p), \underline{u}(t-1), \underline{u}(t-2),...,\underline{u}(t-q),$$
$$\Theta, \underline{\varepsilon}(t), \underline{\varepsilon}((t-1),..., \underline{\varepsilon}(t-r)) \tag{4.4}$$

where p, q, and r represent the orders of past variables in the vectors of $\underline{y}(t)$, $\underline{u}(t)$, and $\underline{\varepsilon}(t)$. For example, for a 2×2 dynamic system, if $\underline{y}(t) = [y_1(t), y_2(t)]^T$, $\underline{u}(t) = [u_1(t), u_2(t)]^T$, and $\underline{\varepsilon}(t) = [\varepsilon_1(t), \varepsilon_2(t)]^T$, then $\underline{y}(t - p) = [y_1(t - p_1), y_2(t - p_2)]^T$, $\underline{u}(t - q) = [u_1(t - q_1), u_2(t - q_2)]^T$, and $\underline{\varepsilon}(t - r) = [\varepsilon_1(t - r_1), \varepsilon_2(t - r_2)]^T$ where p_1 and p_2, q_1 and q_2, and r_1 and r_2 are the maximum orders of the past outputs related to the present outputs $y_1(t)$ and $y_2(t)$, the past inputs related to the present inputs $u_1(t)$ and $u_2(t)$, and the past residuals related to the present residuals $\varepsilon_1(t)$ and $\varepsilon_2(t)$, respectively.

Obviously this model assumption is much more complex than the static case described in Eq. (4.3). In Eq. (4.4), the output vector at current time instant $\underline{y}(t)$ is assumed to relate itself, system input vector $\underline{u}(t)$, and modeling residual vector $\underline{\varepsilon}(t)$ from past one time instant up to past p, q, and r time instants, respectively.

Eq. (4.4) is a general form of discrete time Nonlinear AutoRegressive Moving Average with eXogenous input (NARMAX) relationship. This general equation can be simplified to obtain the following equation of a discrete time Nonlinear AutoRegressive with eXogenous input (NARX) relationship. This relationship has been widely used in process modeling, prediction, and control.

$$\underline{y}(t) = f(\underline{y}(t-1), \underline{y}(t-2), \ldots, \underline{y}(t-p), \underline{u}(t-1), \underline{u}(t-2), \ldots, \underline{u}(t-q), \Theta) + \underline{\varepsilon}(t)$$

$$(4.5)$$

Actually, static models represent dynamic models in steady state, that is, static models describe the relationships between state variables in steady state.

Eqs. (4.4) and (4.5) present a general form of model structure based on input–output data $\{\underline{u}(1), \underline{y}(1), \underline{u}(2), \underline{y}(2), \ldots, \underline{u}(N-1), \underline{y}(N-1), \underline{u}(N), \underline{y}(N)\}$ measured from the system. This method for dynamic system modeling from input–output measurements is called system identification (Ljung, 1999). In general, input–output signals in a system or process are measurable. Because the dynamic characteristics of a process are reflected in the input–output data, the mathematical model of the process can be built using the input–output data. This is what system identification implies.

The mechanism of practical food manufacturing and processing systems is complex. Describing mass and heat transfer activities in complex food manufacturing and processing systems is difficult. Therefore, it is not easy to build mathematical models for these processes theoretically. However, if only the external characteristics of the processes are of concern, these processes can be viewed as a "black box" as shown in Figure 4.1. In terms of the information of inputs and outputs presented to the black box, a process model with external characteristics equivalent to the characteristics of the black box, can be built. This is a direct application of system identification in food quality process modeling.

In some sense, the method of input–output modeling has an advantage over the method of theoretical modeling because it is not necessary to know

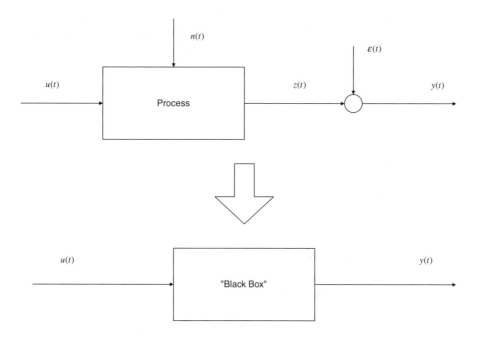

Figure 4.1 "Black box" structure of system identification.

the internal operating mechanism of the process for system identification. The key for success in system identification is to design an effective experiment to produce a maximal amount of information for modeling, but it is very difficult to do so even though various powerful tools, such as PRBS-based statistical experiment design and analysis methods are available. Therefore, in practical dynamic process modeling, two modeling methods, system identification and theoretical, are often used together to achieve the goal of the modeling task. In the theory of modeling, the problems in theoretical modeling are viewed as a "white box"; the problems in system identification modeling are viewed as a "black box"; and the problems combining both theoretical and system identification modeling are viewed as a "gray box," in which the part of known mechanism uses theoretical modeling, and the part of unknown mechanism uses system identification modeling.

The focus of this chapter is on food process modeling using input–output data based on the method of system identification. In general, the procedure for identification of a process input–output model includes the following steps:

1. Identification and preparation—the objective of modeling needs to be determined. This is important for the determination of modeling type, methods, and accuracy. For example, if the model is used for constant control, the requirement for accuracy should not be high; if the model is used for servos or forecasting and prediction, the

requirement for accuracy cannot be low. In addition, before identification, some information needs to be collected or measured from the concerned process in some way, such as nonlinearity, time constant, transient response, noise characteristics, process operating conditions, and so on. The prior knowledge of the process will direct the experiment design.

2. Experimental design—experimental design includes the choice and determination of the following:
   a. Input signal (magnitude, bands, etc.).
   b. Sampling time.
   c. Identification time (length of data).
   d. Open or closed loop identification.
   e. Off-line or on-line identification.

3. Data preprocessing—generally, this involves transform of input–output data into zero-mean sequences and elimination of high frequency components in the input–output data. This work could improve identification accuracy significantly if appropriately done.

4. Model structure identification—this includes a prior assumption of model structure and the determination of model structure parameters. Model structure parameters usually are orders of input–output variables and the amount of time lag between the input variables and the output variables.

5. Model parameter identification—after the model structure is determined, the model parameters need to be identified. Many methods are available for model parameter identification. The Least Squares is the most popular of these methods.

6. Model validation—model validation involves procedures to evaluate how the identified model fits the input–output data in terms of the modeling objectives and prior knowledge of the process. If the model is rejected according to the designated criterion, the previous steps are repeated with appropriate adjustments. If the model is accepted as an acceptable description of the concerned process, the model can be put into practice.

4.2 Linear statistical modeling

Statistical modeling is based on conventional linear regression techniques. The linear regression can be divided further into simple regression which is used to establish the statistical relationship between a single input and a single output as described earlier, and multiple regression, which is used to establish the statistical relationships between multiple inputs and single or multiple outputs. The principle of multiple regression is similar to simple regression, but the computation is much more complex.

For example, consider the relationship between one output and two or more inputs with the technique of multiple regression. If a variable y in a system is correlated linearly with n variables $x_1, x_2,..., x_n$, then the linear regression model between these variables can be expressed as

$$y = \beta_0 + \beta_1 x_1 + \beta_2 x_2 + \cdots + \beta_n x_n + \varepsilon \tag{4.6}$$

In this expression, ε is assumed to have the following properties:

1. $E(\varepsilon) = 0$, that is, the random variable has 0 mean.
2. $\mathrm{Cov}(\varepsilon_i, \varepsilon_j) = 0 \, (i \neq j)$ and $\sigma_\varepsilon^2 \, (i = j)$ where $\mathrm{Cov}(\varepsilon_i, \varepsilon_j)$ represents the covariance of ε_i and ε_j, that is, each observation is independent.
3. The variables $x_i \, (i = 1, 2, \ldots, n)$ are nonrandom and determinant, that is, $E(\varepsilon x_i) = 0$.

It is impossible to know the real value of the population, so the output cannot be evaluated from the population regression equation

$$E(y) = \beta_0 + \beta_1 x_1 + \beta_2 x_2 + \cdots + \beta_n x_n \tag{4.6a}$$

The sample regression model is

$$E(y) = \hat{\beta}_0 + \hat{\beta}_1 x_1 + \hat{\beta}_2 x_2 + \cdots + \hat{\beta}_n x_n + e \tag{4.7}$$

where e is the error term of the model.

The population is inferred from this sample regression model as follows

$$\hat{y} = \hat{\beta}_0 + \hat{\beta}_1 x_1 + \hat{\beta}_2 x_2 + \cdots + \hat{\beta}_n x_n \tag{4.8}$$

The work of multiple linear regression is to calculate the regression coefficients $\hat{\beta}_i \, (i = 1, 2, \ldots, n)$ based on sample observations $(y_i, x_{i1}, x_{i2}, \ldots, x_{in})$ $(i = 1, 2, \ldots, N)$, and to have these coefficients tested statistically to show the confidence levels of the estimates.

Now let us discuss the calculation of the regression coefficients. Least squares is the most widely used method for calculation of the regression coefficients. Calculation with the least squares is a type of optimization to produce the coefficients to make the model "best" fit the data. From Eqs. (4.6) and (4.8), there exists

$$\begin{aligned}
Q &= \sum_{i=1}^{N} e_i^2 \\
&= \sum_{i=1}^{N} (y_i - \hat{y}_i)^2 \\
&= \sum_{i=1}^{N} [y_i - (\hat{\beta}_0 + \hat{\beta}_1 x_1 + \hat{\beta}_2 x_2 + \cdots + \hat{\beta}_n x_n)]^2
\end{aligned} \tag{4.9}$$

For the given data, Eq. (4.9) represents that Q is a nonnegative sum of squares of deviations of these regression coefficients $\hat{\beta}_1, \hat{\beta}_2, \ldots, \hat{\beta}_n$, so there must exist a minimal value for Q. In terms of the principle of extremum in calculus, the following simultaneous equations hold for $\hat{\beta}_1, \hat{\beta}_2, \ldots, \hat{\beta}_n$

$$
\begin{cases}
\dfrac{\partial Q}{\partial \hat{\beta}_0} = 0 \\[2ex]
\dfrac{\partial Q}{\partial \hat{\beta}_1} = 0 \\[1ex]
\vdots \\[1ex]
\dfrac{\partial Q}{\partial \hat{\beta}_n} = 0
\end{cases}
\tag{4.10}
$$

It can be derived from $\partial Q / \partial \hat{\beta}_0 = 0$ that

$$
2 \sum_{i=1}^{N} [y_i - (\hat{\beta}_0 + \hat{\beta}_1 x_{i1} + \cdots + \hat{\beta}_n x_{in})](-1) = 0
\tag{4.11}
$$

If the sample means of the observations are expressed as

$$
\begin{cases}
\bar{x}_j = \dfrac{1}{N} \sum_{i=1}^{N} x_{ij} (j = 1, 2, \ldots, n) \\[2ex]
\bar{y} = \dfrac{1}{N} \sum_{i=1}^{N} y_i
\end{cases}
$$

Then, Eq. (4.11) becomes

$$
\hat{\beta}_0 = \bar{y} - (\hat{\beta}_1 \bar{x}_1 + \hat{\beta}_2 \bar{x}_2 + \cdots + \hat{\beta}_n \bar{x}_n)
\tag{4.12}
$$

Replace this equation in Eq. (4.8) with

$$
\begin{aligned}
X_{ij} &= x_{ij} - \bar{x}_j \\
Y_i &= y_i - \bar{y} \qquad (j = 1, 2, \ldots, n; i = 1, 2, \ldots, N)
\end{aligned}
$$

So,

$$
\hat{y}_i = \bar{y} + \hat{\beta}_1 X_{i1} + \hat{\beta}_2 X_{i2} + \cdots + \hat{\beta}_n X_{in}
\tag{4.13}
$$

and

$$
Q = \sum_{i=1}^{N} [Y_i - (\hat{\beta}_1 X_{i1} + \hat{\beta}_2 X_{i2} + \cdots + \hat{\beta}_n X_{in})]^2
\tag{4.14}
$$

Now except $\partial Q/\partial\hat{\beta}_0$, n other equations in the simultaneous Eqs. (4.10) become

$$\begin{cases} \dfrac{\partial Q}{\partial\hat{\beta}_1} = 2\sum_{i=1}^{N}[Y_i - (\hat{\beta}_1 X_{i1} + \hat{\beta}_2 X_{i2} + \cdots + \hat{\beta}_n X_{in})](-X_{i1}) = 0 \\[2mm] \dfrac{\partial Q}{\partial\hat{\beta}_2} = 2\sum_{i=1}^{N}[Y_i - (\hat{\beta}_1 X_{i1} + \hat{\beta}_2 X_{i2} + \cdots + \hat{\beta}_n X_{in})](-X_{i2}) = 0 \\[2mm] \dfrac{\partial Q}{\partial\hat{\beta}_n} = 2\sum_{i=1}^{N}[Y_i - (\hat{\beta}_1 X_{i1} + \hat{\beta}_2 X_{i2} + \cdots + \hat{\beta}_n X_{in})](-X_n) = 0 \end{cases} \qquad (4.15)$$

Rearrange these equations with

$$\begin{cases} s_{jl} = s_{lj} = \sum_{i=1}^{N} X_{ij} X_{il} \\[2mm] s_{jy} = \sum_{i=1}^{N} X_{ij} Y_i \end{cases} \qquad (j, l = 1, 2, \ldots, n) \qquad (4.16)$$

Then, Eq. (4.13) becomes

$$\begin{cases} s_{11}\hat{\beta}_1 + s_{12}\hat{\beta}_2 + \cdots + s_{1n}\hat{\beta}_n = s_{1y} \\ s_{21}\hat{\beta}_1 + s_{22}\hat{\beta}_2 + \cdots + s_{2n}\hat{\beta}_n = s_{2y} \\ \vdots \\ s_{n1}\hat{\beta}_1 + s_{n2}\hat{\beta}_2 + \cdots + s_{nn}\hat{\beta}_n = s_{ny} \end{cases} \qquad (4.17)$$

These are normal equations for multivariate least squares. When given the observation data, s_{jl} and s_{jy} can be solved iteratively. Therefore, the normal equations, Eqs. (4.17), are the nth order linear simultaneous equations with $\hat{\beta}_1$, $\hat{\beta}_2$, ..., $\hat{\beta}_n$ as the unknown variables. If the number of samples N is greater than the number of unknown variables n, that is, $N > n$, and any unknown variable is not a linear combination of other unknown variables, then Eqs. (4.17) have a unique solution which can be represented as

$$\hat{\beta}_j = \sum_{i=1}^{n} c_{jl} s_{ly} \qquad (j = 1, 2, \ldots, n) \qquad (4.18)$$

where c_{jl} is the element of the matrix $(c_{jl}) = (s_{jl})^{-1}$ $(j, l = 1, 2, \ldots, n)$. Once $\hat{\beta}_j$ $(j = 1, 2, \ldots, n)$ are solved, they can be used to solve $\hat{\beta}_0$ with Eq. (4.12).

The calculation for multiple regression coefficients is similar to the one for simple regression coefficients, but the amount of computation is significantly more. For the model of simple regression represented as Eqs. (4.1) and (4.2), the regression coefficients $\hat{a}$ and $\hat{b}$ can be calculated as

$$\begin{cases} \hat{b} = \dfrac{\sum_{i=1}^{N} X_i Y_i}{\sum_{i=1}^{N} X_i^2} \\ \hat{a} = \bar{y} - \hat{b}\bar{x} \end{cases} \tag{4.19}$$

It can be proven that the least squares estimate $\hat{\beta}_j$ of the multiple linear regression and $\hat{a}$ and $\hat{b}$ of the simple linear regression are linear, unbiased, and with minimal variance (Milton and Arnold, 1990). They are linear, unbiased, and minimal variance estimates of the population variables β_j and a and b.

When the coefficients of the sample regression equation are estimated, it is necessary to perform statistical tests in order to give the confidence level for population inference from the equation.

If used just for parameter estimation, no assumption on the distribution of ε is needed. However, if the issues of statistical tests and confidence interval are concerned, it is necessary to assume the distribution of ε (or the distribution of $\hat{\beta}_j$). In general, there are two types of distribution assumptions on ε

1. Assume ε is normally distributed. This is a strict assumption.
2. Do not assume ε to be specifically distributed. However, in terms of the central limit theorem (Milton and Arnold, 1990), when the sample size tends to be large, the distribution of ε is approximately normally distributed.

Considering the preceding assumptions, the tests of significance include the significance test of each independent variable and the overall significance test of all independent variables. The following hypothesis is often constructed for the significance test of each independent variable

$$H_0: \beta_j = 0 \quad (j = 1, 2, \ldots, n)$$
$$H_1: \beta_j \neq 0 \quad (j = 1, 2, \ldots, n)$$

where H_0 is the null hypothesis and H_1 is the alternative hypothesis. For the case of ε being assumed to be normally distributed, y is normally distributed as well, and β_j is the combination of y_j. Therefore,

$$\hat{\beta}_j \sim N\left(\beta_j, \sqrt{\text{Var}(\hat{\beta}_j)}\right)$$

Using the Z transform, we have

$$Z = \frac{\hat{\beta}_j - \beta_j}{\sqrt{\text{Var}(\hat{\beta}_j)}} \sim N(0, 1) \qquad (4.20)$$

When a level of significance, α, is given, the value of $N_{\alpha/2}$ can be looked up in the table of standard normal distribution (refer to any book about basic statistics, e.g., Milton and Arnold, 1990). The two-tailed test with the significance level α, if $|Z| < |N_{\alpha/2}|$, means that $\hat{\beta}_j$ is not significantly different from 0, that is, $\hat{\beta}_j$ and β_j are not significantly different. In this case, the null hypothesis H_0 is accepted. Otherwise, $|Z| > |N_{\alpha/2}|$, and it means that $\hat{\beta}_j$ is significantly different from 0, that is, $\hat{\beta}_j$ and β_j are significantly different. In this case, the null hypothesis H_0 is rejected, and the alternative hypothesis H_1 is accepted.

The preceding test based on the Z statistic is for problems with a large sample size (number of samples greater than 30). For the problems with a small sample size (number of samples less than 30), the student t statistic is often used. The student t statistic depends on the degree of freedom $N - n$ and uses the variance estimate to replace the true variance of the population

$$t = \frac{\hat{\beta}_j - \beta_j}{\sqrt{\widehat{\text{Var}}(\hat{\beta}_j)}} \qquad (4.21)$$

When performing a test with the student t statistic with the given level of significance α, the value of $t_{\alpha/2}$ can be looked up in the table for the student t distribution for the degree of freedom $N - n$. If $|t| < |t_{\alpha/2}|$, $\hat{\beta}_j$ and β_j are not significantly different and the null hypothesis H_0 is accepted; otherwise, if $|t| > |t_{\alpha/2}|$, $\hat{\beta}_j$ and β_j are significantly different, the null hypothesis H_0 is rejected and the alternative hypothesis H_1 is accepted.

The overall significance test is for the test of the significance of n independent variables $x_1, x_2, \ldots, x_n$ on y. For this, the following hypothesis needs to be constructed

$$H_0: \beta_1 = \beta_2 = \ldots = \beta_n = 0$$
$$H_1: \text{not all } \beta_j = 0, \quad j = 1, 2, \ldots, n$$

This implies that if H_0 is accepted, all of the population parameters are zero and the linear relationship between y and these independent variables does not exist. Otherwise, if H_0 is rejected, then it can be inferred that the n independent variables have significant impact on y.

As described earlier, when any of the independent variables cannot be represented by other independent variables, the normal Eqs. (4.17) have a unique solution. Under this condition, the following F statistic observes the F distribution with the degree of freedom of n and $N - n - 1$

$$F = \frac{\sum_{i=1}^{N}(y_i - \hat{y}_i)^2/n}{\sum_{i=1}^{N}(\hat{y}_i - \bar{y})^2/(N - n - 1)} \sim F(n, N - n - 1) \qquad (4.22)$$

When given the level of significance α, the value of F_α can be looked up in the table of F distribution according to the degree of freedom $v_1 = n$ and $v_2 = N - n - 1$. If $F > F_\alpha$, then reject H_0 and accept H_1, which means that these independent variables have significant impact on y. Otherwise, if $F < F_\alpha$, then accept H_0, which means that none of these independent variables has a significant impact on y.

Further, the F test can be extended to test the significance of some of the regression coefficients. Variable selection is very useful in practice. Assume y strongly impacted by n independent variables $x_1, x_2, ..., x_n$. The regression model can be expressed as Eq. (4.6). If it is also impacted by other $l - n$ independent variables $x_{n+1}, x_{n+2}, ..., x_l$ $(l > n)$, the regression model, including all independent variables, should be

$$y = \beta_0 + \beta_1 x_1 + \beta_2 x_2 + \cdots + \beta_n x_n + \beta_{n+1} x_{n+1} + \beta_{n+2} x_{n+2} + \cdots + \beta_l x_l + \varepsilon \qquad (4.23)$$

Now, consider the impact of each of $x_{n+1}, x_{n+2}, ..., x_l$ on y, and at the same time consider whether they all have the impact on y. The former problem can be solved using the Z or t test as described previously. Here the latter problem only will be discussed. For this, the following hypothesis can be constructed

$$H_0: \beta_{n+1} = \beta_{n+2} = \cdots = \beta_l = 0$$
$$H_1: \text{not all } \beta_j = 0, \quad j = n + 1, n + 2, ..., l$$

Define the following terms

$$SST = \sum_{i=1}^{N}(y_i - \bar{y})^2, \quad SSR = \sum_{i=1}^{N}(y_i - \hat{y}_i)^2, \quad \text{and} \quad SSE = \sum_{i=1}^{N}(\hat{y}_i - \bar{y})^2$$

It can be proved that for Eq. (4.6), there exists

$$SST_n = SSR_n + SSE_n$$

and for Eq. (4.23), there exists

$$SST_l = SSR_l + SSE_l$$

So, the F statistic for the increase of new independent variables in Eq. (4.23) with respect to Eq. (4.6) is

$$F = \frac{(SSR_l - SSR_n)/(l - n)}{SSE_l/(N - l - 1)} \qquad (4.24)$$

Calculate the value of F in Eq. (4.24) and compare it with the value in the table of F distribution for the degree of freedom $l - n$ and $N - l - 1$. By doing so, H_0 is either rejected or accepted, and the impact of some of the regression coefficients in independent variables on y can be inferred.

Because simple regression has only one independent variable, there is no need to test the overall impact. The impact of each regression coefficient is necessary. For the case of a large sample size (number of samples greater than 30), the Z statistic for a and b is, respectively,

$$
\begin{cases}
Z_a = \dfrac{\hat{a} - a}{\sqrt{\sigma_\varepsilon^2 \sum\limits_{i=1}^{N} x_i^2 / \sum\limits_{i=1}^{N} X_i^2}} \sim N(0, 1) \\[4ex]
Z_b = \dfrac{\hat{b} - b}{\sqrt{\sigma_\varepsilon^2 / \sum\limits_{i=1}^{N} X_i^2}} \sim N(0, 1)
\end{cases}
\tag{4.25}
$$

So, with the given level of significance, the decision will be made to accept or reject the null hypothesis based on the value in the table of standard normal distribution.

For the case of a small sample size (number of samples less than 30), the t statistic for a and b is, respectively,

$$
\begin{cases}
t_a = \dfrac{\hat{a} - a}{\sqrt{\sigma_\varepsilon^2 \sum\limits_{i=1}^{N} X_i^2 / \left(N \sum\limits_{i=1}^{N} X_i^2 \right)}} \\[4ex]
t_b = \dfrac{\hat{b} - b}{\sqrt{\sigma_\varepsilon^2 / \sum\limits_{i=1}^{N} X_i^2}}
\end{cases}
\tag{4.26}
$$

Similarly, with the given level of significance, the decision will be made to accept or reject the null hypothesis based on the value in the table of t distribution with the degree of freedom $N - 1$.

Instead of t and F tests, the reliability of a regression equation is frequently measured by the multiple correlation coefficient. The multiple correlation coefficient can be represented as $R_{y \cdot x_1, x_2, \ldots, x_n}$ or R for short. This parameter measures the correlation between y and all the independent variables $x_1, x_2, \ldots, x_n$. It can be interpreted as the proportion of the variability that has been accounted for by the regression equation, that is,

$$
R^2 = \frac{\sum\limits_{i=1}^{N} \hat{Y}_i^2}{\sum\limits_{i=1}^{N} Y_i^2} = \frac{\sum\limits_{i=1}^{N} (\hat{y}_i - \bar{y})^2}{\sum\limits_{i=1}^{N} (y_i - \bar{y})^2} = 1 - \frac{\sum\limits_{i=1}^{N} (y_i - \hat{y}_i)^2}{\sum\limits_{i=1}^{N} (y_i - \bar{y}_i)^2}
\tag{4.27}
$$

R^2 has another name, the coefficient of multiple determination. If $R^2 = 1$, the model is perfect. If $R^2 = 0$, the model does no better than predicting the average value it tries to estimate.

In statistical modeling, often a subset of the full set of independent variables can contribute to build a model to adequately fit the data. The model with the full set of independent variables may increase the complexity of the standard errors of the estimates and add little accuracy to the estimates. In general, there are two approaches commonly used to select a subset of the possible independent variables to build the smallest and most adequate model: backward elimination and forward addition. Backward elimination is a step-down procedure. With this procedure, the process of modeling begins with the model with the full set of independent variables. The variables, then, are eliminated one at a time as long as they are determined to make little contribution to the model. The approach of forward addition is a step-up procedure. In this procedure, the model is built by adding independent variables one at a time through testing their contribution to the model. This approach was used to establish the method of stepwise regression (Efroymson, 1962). Stepwise regression tests the contribution of each variable at each step. A variable that is already in the model will be removed later when its contribution is redundant with some other variables.

For structure identification of process models, the values of the orders of output $y(t)$ and input $u(t)$, numbers of past $y(t)$ and $u(t)$ used in model input, are determined. Typically, as the orders of $y(t)$ and $u(t)$ increase, the accuracy of the model increases. However, if the order is too large, the model will overfit the data, that is, the model will fit the data extremely well but poorly predict outputs based on data not included in building the model. In order to balance the model accuracy against the order of the model, Akaike's information criterion (AIC) (Akaike, 1973) can be used. Order determination with AIC is an application of the principle of maximum likelihood. It looks for a model where the output probability distribution approximates the probability distribution of the actual process output to the maximum degree. AIC is defined as

$$\text{AIC}(N_0) = -2 \log(ml) + 2N_0 \tag{4.28}$$

where ml is the maximum likelihood function of the output of the process model, and N_0 is the total number of parameters in the model. In AIC, the term $-2 \log(ml)$ measures the error of the model, and the term $2N_0$ provides a penalty for the complexity of the model, which penalizes the model with a large N_0. A high value of model error rejects the overly simple model, and a high value of penalty term rejects the highly complex model. Therefore, a model that achieves the least AIC will produce optimal model orders. Because the penalty term increases linearly with the number of parameters in the model and the model error term decreases with each additional parameter, a minimum point of AIC is guaranteed. Figure 4.2 shows the relationship between $\text{AIC}(N_0)$ and N_0.

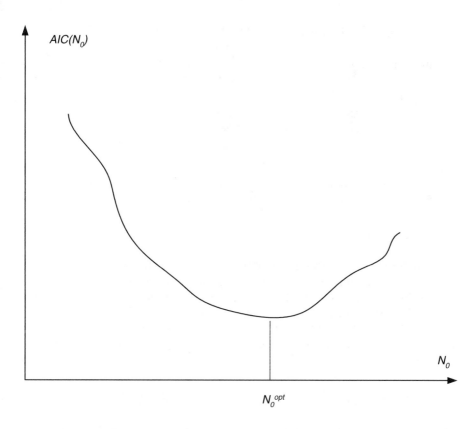

Figure 4.2 Relationship between AIC(N_0) and N_0 where N_0^{opt} is the optimal value of N_0.

4.2.1 Example: Linear statistical modeling based on ultrasonic A-mode signals for beef grading

Whittaker et al. (1992) and Park (1991) conducted regression analysis to obtain a functional relationship between ultrasonic A-mode parameters and the beef tissue constituent property. The first linear regression model was developed to estimate fat concentration from a longitudinal in the time domain. This simple regression model was built based on 100 randomly selected samples among 124 meat samples. The model used to predict intramuscular fat concentration through ultrasonic speed was

$$\% \text{ Fat } = 330.86 - 0.2114 \times \text{LSPD}$$
$$R^2 = 0.66$$

where LSPD is longitudinal speed (m/sec).

Table 4.1 Multiple Regression Models of Different Ultrasonic Probes and Corresponding R^2 Values*

Probe (MHz)	R^2	Model
Longitudinal		
1	0.37	% Fat = 3.392 + 7.078f_b − 0.143 B^* + 1.293f_{sk} − 0.955 Lm
2.25	0.41	% Fat = 5.412 − 1.262f_a − 0.055 B^* + 0.98 Lm
5	0.55	% Fat = 0.733 − 8.937f_b + 8.507f_c + 1.591 Lm
Shear		
1	0.50	% Fat = 0.331 − 0.059 B^* + 0.784f_{sk} + 0.731 Lm
2.25	0.82	% Fat = 1.790 − 2.373f_a + 0.049 B^* + 1.178 Lm
5	0.58	% Fat = −2.539 + 1.148f_c + 0.864 Lm

* Adapted from Park, (1991). With permission.

Further, based on the frequency analysis data, the multiple regression models of six different ultrasonic probes were built for the prediction of fat concentration. The result showed that the determinant coefficients, R^2, of the models of the shear probes were greater than those for the longitudinal probes. Table 4.1 shows the models of different ultrasonic probes and the corresponding values of R^2. The full model should be in the form

$$\% \text{ Fat} = \hat{\beta}_0 + \hat{\beta}_1 f_a + \hat{\beta}_2 f_b + \hat{\beta}_3 f_p + \hat{\beta}_4 f_c + \hat{\beta}_5 B^* + \hat{\beta}_6 f_{sk} + \hat{\beta}_7 Lm$$

Some of the variables do not appear in the table. This indicates that the corresponding regression coefficients are zero, and these variables do not strongly relate to the output. From the table, it can be seen that the highest value of R^2 occurred for the model of the 2.25 MHz shear probe and that the local maxima Lm appeared as the dominant term in all models.

4.2.2   Example: Linear statistical modeling for food odor pattern recognition by an electronic nose

The purpose of food odor pattern recognition by an electronic nose is to classify the food samples based on readings of a sensor array in an electronic nose. Linear statistical modeling for food odor pattern recognition by an electronic nose found the following relationship

$$\hat{c} = \hat{\beta}_0 + \hat{\beta}_1 \tilde{x}_1 + \hat{\beta}_2 \tilde{x}_2 + \cdots + \hat{\beta}_n \tilde{x}_n \tag{4.29}$$

where $\hat{c}$ is the sample classification assignment, $\tilde{x}_i (i = 1, 2, \ldots, n)$ is the ith normalized reading in the n-sensor array, and $\hat{\beta}_i (i = 0, 1, 2, \ldots, n)$ is the ith coefficient estimate of the corresponding sensor reading.

With a 32-sensor commercial electronic nose (AromaScan), a classification model of peanut off-flavor had a binary output of 1 representing off-flavor and

0 representing non-off-flavor, related 32 inputs to the binary output (Osborn et al., in press).

4.2.3   Example: Linear statistical modeling for meat attribute prediction based on textural features extracted from ultrasonic elastograms

Lozano (1995), Moore (1996), and Huang et al. (1997) conducted statistical modeling for meat attribute prediction based on textural features extracted from ultrasonic elastograms in a variety of data sets. Their work was the prediction of mechanical and chemical attributes of beef samples of LD muscle. One original and one replicate elastogram were generated from each beef sample and were averaged to produce a single image. The number of samples was 29. A total of 10 mechanical and chemical attributes data of the beef samples were collected.

Haralick's statistical textural features and wavelet textural features were extracted from the elastograms of the beef samples (Huang et al., 1997).

For a single elastogram, if all 14 parameters originally presented by Haralick et al. (1973) are computed for 4 angles (0°, 45°, 90°, and 135°) and 4 neighborhood distances (d = 1, 2, 5, and 10), there will be 224 (4 × 4 × 14) independent statistical textural features. In this application, the feature values in angles were averaged. In this way, there still existed 56 (4 × 14) independent statistical textural feature parameters.

In using the method of regular regression, if all of the 54 independent statistical textural feature parameters are used as the input, a unique solution for the regression equation will not exist because the number of parameters (55) is greater than the number of the samples (29). So, the linear regression models were fitted by separating distances at 1, 2, 5, and 10. In this way, each model had 14 inputs. Table 4.2 shows the values of R^2 of all models.

Table 4.2 R^2 Values of Linear Regression Models for Beef Attribute Prediction Based on Haralick's Statistical Textural Feature Parameters*

R^2	Distance 1 Average Angle	Distance 2 Average Angle	Distance 5 Average Angle	Distance 10 Average Angle
WB1 (kg)	0.3894	0.2868	0.3997	0.5597
WB2 (kg)	0.4077	0.4323	0.4417	0.5998
WB3 (kg)	0.6808	0.7629	0.7700	0.5774
WB4 (kg)	0.3580	0.4552	0.2917	0.4222
Calp (μg/g)	0.2591	0.1701	0.2299	0.1898
Sarc (μm)	0.4226	0.4574	0.5158	0.5079
T.Coll (mg/g)	0.3569	0.3873	0.3875	0.4060
%Sol	0.5781	0.5804	0.3687	0.4961
%Mois	0.4769	0.3564	0.4969	0.5248
%Fat	0.4687	0.7423	0.6193	0.6150

* Adapted from Huang et al. (1997). With permission.

In order to find major independent variables and remove minor independent variables, the method of stepwise regression was used with a level of significance of $\alpha < 0.15$ to determine the effectiveness of the information extracted from elastograms (Haralick's statistical textural features) to predict WB1, WB2, WB3, WB4, Calp, Sarc, T.Coll, %Sol, %Mois, and %Fat. The final multiple regression equations were

$$WB1 = -2.06 - 254.90 f_{14@d5} - 45.60 f_{12@d2}$$
$$+ 0.4 f_{2@d1} + 0.2 f_{2@d2} - 0.2 f_{2@d10} \qquad (R^2 = 0.47)$$
$$WB2 = -38.62 - 177.95 f_{12@d1} + 235.56 f_{12@d2}$$
$$+ 49.74 f_{13@d2} \qquad (R^2 = 0.40)$$
$$WB3 = 2.20 - 14.83 f_{12@d2} - 124.24 f_{14@d5} \qquad (R^2 = 0.40)$$
$$\%Sol = 16.36 + 0.01 f_{2@d5} \qquad (R^2 = 0.12)$$
$$\%Mois = 41.43 + 41.21 f_{13@d1} \qquad (R^2 = 0.23)$$
$$\%Fat = 16.15 - 5.89 f_{8@d10} \qquad (R^2 = 0.25)$$

However, the equations of WB4, Calp, Sarc, T.Coll could not be established because the operation of stepwise regression failed in building models for them owing to the lack of a significance for the parameter test. In the equation of WB1, $f_{14@d5}$ is the 14th Haralick's statistical textural feature, maximal correlation coefficient, at the distance of 5, $f_{12@d2}$ is the 12th Haralick's statistical textural feature; informational measure of correlation −1, at the distance of 2, $f_{2@d1}$ is the second Haralick's statistical textural feature; contrast, at the distance of 1, $f_{2@d2}$ is also the second Haralick's statistical textural feature but at the distance of 2; and $f_{2@d10}$ is the second Haralick's statistical textural feature at the distance of 10.   In this equation, the remaining 49 independent variables were removed by the process of stepwise regression because they did not significantly impact on the dependent variable WB1. The equations of WB2, WB3, %Sol, %Mois, and %Fat were built in a similar way.

All of the elastograms of beef samples were cropped into the size of 128×128 for wavelet analysis. The number of independent wavelet textural features was 29 ($4 \times 7 + 1$). In this way, if all of the 29 feature parameters were used, fewer samples would be used to determine more parameters because the number of the samples was 29. Therefore, the wavelet decomposition was performed one step less than the standard procedure, and 25 feature parameters were produced instead. Table 4.3 shows that all regression models based on wavelet textural feature parameters had much higher R^2 values than the models based on Haralick's statistical textural feature parameters. This indicates that these models account for a high percentage of the variations in the outputs as beef mechanical and chemical attributes and that the wavelet textural features can be used to effectively predict these beef attributes.

Table 4.3 R^2 Values of Linear Regression Models for Beef Attribute Prediction Based on Wavelet Textural Feature Parameters*

R^2	Daubechies-4 Wavelet Textural Features
WB1 (kg)	0.9562
WB2 (kg)	0.9432
WB3 (kg)	0.9008
WB4 (kg)	0.8392
Calp (μg/g)	0.8349
Sarc (μm)	0.8770
T.Coll (mg/g)	0.8798
%Sol	0.9060
%Mois	0.9100
%Fat	0.8851

* Adapted from Huang et al. (1997). With permission.

Figure 4.3 Black box structure of the snack food frying process.

4.2.4 Example: Linear statistical dynamic modeling for snack food frying process control

Based on the black box idea, the snack food frying process can be seen in Figure 4.3. Mathematically, Eq. (4.5) can be simplified further to represent the snack food frying process in the assumption of a linear SISO system:

$$y(t) = \alpha_1 y(t-1) + \alpha_2 y(t-2) + \cdots + \alpha_p y(t-p) + \beta_1 u(t-d-1)$$
$$+ \beta_2 u(t-d-2) + \cdots + \beta_q u(t-d-q) + \varepsilon(t) \tag{4.30}$$

where d represents the time lag between the input and output. Comparison of Eqs. (4.29) and (4.5) shows the latter to be significantly simplified. Eq. (4.30) is a discrete time AutoRegressive with eXogenous input (ARX) description of the process. In general, for linear systems the models can be described in this form as long as $y(t)$ and $h(t)$ are observable at discrete instants:

$$y(t) = \underline{h}^T(t) \cdot \Theta + \varepsilon(t) \tag{4.31}$$

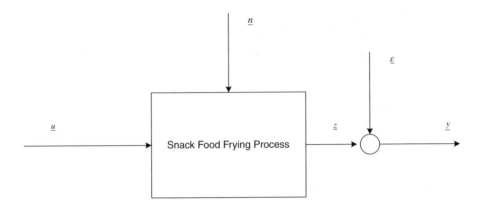

Figure 4.4 A process structure to be identified.

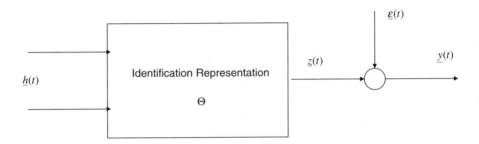

Figure 4.5 Representative structure of an identified model.

where $\underline{h}(t) = (h_1(t), h_2(t), \ldots, h_n(t))^T$ is the model input vector, and $\Theta = (\theta_1, \theta_2, \ldots, \theta_n)^T$ is the vector for model unknown parameters. This is the so-called least squares form of the model. From this form, it is not difficult to get the least squares solution of the model.

If $\underline{h}(t) = (y(t-1), y(t-2), \ldots, y(t-p), u(t-d-1), u(t-d-2), \ldots, u\,(t-d-q))^T$ and $\Theta = (\alpha_1, \alpha_2, \ldots, \alpha_p, \beta_1, \beta_2, \ldots, \beta_q)^T$, then Eq. (4.5) can be written in the form of Eq. (4.30). This is described in Figures 4.4 and 4.5. For the process to be identified as shown in Figure 4.3, the model describing it can be expressed in the form shown as Figure 4.4. This is a form of least squares. It should be noted that in this form the output scalar $y(t)$ is the linear combination of input vector $\underline{h}(t)$, and the input $\underline{h}(t)$ is no longer just $u(t)$ which includes the past inputs of the original process, $u(t-\cdot)$, and past outputs of the original process, $y(t-\cdot)$ and at the same time, model residual $\varepsilon(t)$ is no longer the original measurement noise $w(t)$.

The preceding model can be extended to the case of MIMO processes as follows

$$\underline{y}(t) = H(t) \cdot \Theta + \underline{\varepsilon}(t) \tag{4.32}$$

where the output vector is

$$\underline{y}(t) = [y_1(t), y_2(t), \ldots, y_m(t)]^T$$

The noise vector is

$$\underline{\varepsilon}(t) = [\varepsilon_1(t), \varepsilon_2(t), \ldots, \varepsilon_m(t)]^T$$

The parameter vector is

$$\Theta = (\theta_1, \theta_2, \ldots, \theta_n]^T$$

The input data matrix is

$$H(t) = \begin{bmatrix} h_{11}(t) & h_{12}(t) & \cdots & h_{1n}(t) \\ h_{21}(t) & h_{22}(t) & \cdots & h_{2n}(t) \\ \vdots & & & \\ h_{m1}(t) & h_{m2}(t) & \cdots & h_{mn}(t) \end{bmatrix}$$

Bullock (1995) used the System Identification Toolbox of MATLAB (The MathWorks, Inc., Natick, MA) to develop two linear multiple-input, single-output ARX models for the snack food frying process. The data were broken into training and validation groups. The training data were used for modeling while the validation data were used to test the predictive ability of the model. First, all the possible reasonable ARX model structures were created using the struc command that produced a matrix holding all possible combinations of orders and lags. Next, the loss functions for all the possible ARX model structures defined in the matrix were calculated using the arxstruc command. Then, the best model structure was determined using the selstruc command with AIC. The last step was to use the arx command to build the ARX model, that is, to estimate the parameters. The model was simulated using the idsim command.

The numbers of past outputs and inputs and time-lags determined for the model structure are shown in Table 4.4. The coefficients for the ARX models are plotted in Figures 4.6 and 4.7. The performance of the model was evaluated by its ability of prediction and will be discussed in the next chapter.

Table 4.4 Model Structural Parameters of the ARX Model*

Variable	Model Related	Time-Lag (5 s)	Order (5 s)
Inlet temperature $u_1(t)$	Input	19	12
Exposure time $u_2(t)$	Input	14	13
Color $y_1(t)$	Output	0	12
Moisture $y_2(t)$	Output	0	12

* Adapted from Bullock, (1995). With permission.

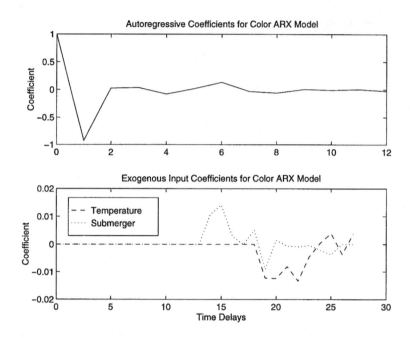

Figure 4.6 Color ARX model coefficients. (From Bullock, 1995. With permission.)

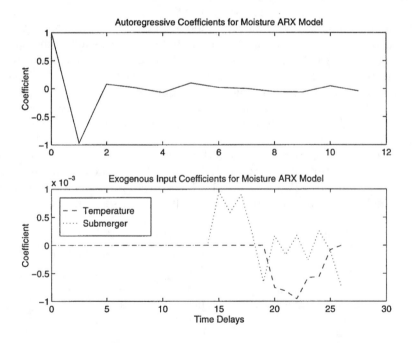

Figure 4.7 Moisture ARX model coefficients. (From Bullock, 1995. With permission.)

4.3   ANN modeling

Artificial neural networks provide a way to organize synthetic neurons to solve complex problems in the same manner as the human brain. Based on studies about the mechanisms and structure of human brain, they have been used as a new computational technology for solving complex problems like pattern recognition, fast information processing, and adaptation. The architecture and implementation of an artificial neural network (ANN) model are a simplified version of the structure and activities of the human brain. For problem solving, the human brain uses a web of interconnected processing units called *neurons* to process information. Each of the neurons is autonomous, independent, and works asynchronously. The vast processing power inherent in biological neural structures has inspired the study of the structure itself as a model for designing and organizing man-made computing structures. McCulloch and Pitts (1943), Hebb (1949), Rosenblatt (1958), Minsky and Papert (1969), Grossberg (1976), and Hopfield (1982) conducted pioneer studies on the theoretical aspect of ANNs. In 1986, the Parallel Distributed Processing (PDP) group published a series of results and algorithms (Rumelhart and McClelland, 1986, Rumelhart et al. 1986a) about back propagation (BP) training for multilayer feedforward networks. This work gave a strong impetus to the area and provided the catalyst for much of the subsequent research and application of ANNs.

An ANN consists of interconnected processing units similar to the neurons in the human brain. It is typically implemented by performing independent computations in some of the units and passing the results of the computations to other units. Each of the processing units performs its computation based on a weighted sum of its input. In an ANN, the processing units are grouped by layers. The input units are grouped as the input layer and the output units are grouped as the output layer. Other units are grouped into hidden layers between the input and the output layer. An activation function usually is used to determine the output of each unit in the hidden and output layers. The connections between processing units, like synapses between neurons, are quantified as weights.

ANNs are used in mathematical modeling to establish a map between system inputs and outputs. ANNs are especially useful when classical statistical modeling, which is based on linear model structure and parameter estimation, cannot be validated. Unlike statistical modeling, before using ANNs in modeling it is not necessary to assume how the system inputs and outputs are related. ANNs always build a relationship between system inputs and outputs as long as they are related in some way in reality. So, ANNs play an important role in applications where classical statistical modeling does not work well.

ANNs have been widely used in nonlinear data modeling, classification, prediction, and process control. In recent years, they have been developed for problem solving in food science analysis and engineering development, including quality quantization and process control.

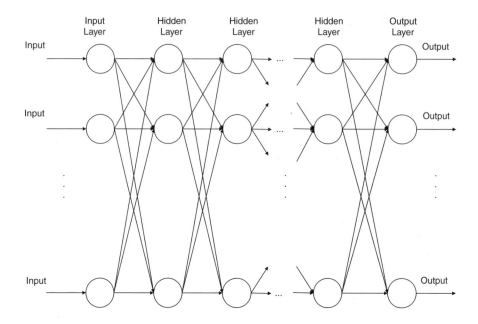

Figure 4.8 Structure of a fully connected MFN.

Review of the contributions made by pioneer and current studies reveal that there exist a number of different ANN architectures. Among them, the multilayer feedforward network with the BP training algorithm is the most widely used in scientific research and engineering applications. A multilayer feedforward network (MFN) includes an input layer with a number of input processing units or nodes, hidden layers with hidden processing units or nodes between the input and output layers, and an output layer with a number of output processing units or nodes. In the network, each input node in the input layer is connected to every node in the adjacent hidden layer. Each node in this hidden layer is connected to every node in the next hidden layer. This is repeated until every node in the last hidden layer is connected to each node in the output layer. The connections are represented by the values of the weights for the strength of the connections. This results in a fully connected MFN. Figure 4.8 shows the structure. Usually, the output of each hidden node is a sigmoid transfer function of the weighted sum on all inputs and a bias term to that node, and the output of each output node in the network is either a sigmoidal function of the summation or a linear function of the summation. This type of MFN, especially with one hidden layer, was proven as the universal approximator for any continuous function (Hornik et al., 1989; Cybenko, 1989).

The training algorithm of the network is as important as the ANN architecture. Consider the one-hidden-layered feedforward neural networks with multiple inputs and single output, the most widely used MFN. The equations for the computation in forward pass are

$$\hat{y} = S_o\left(w_0 + \sum_{i=1}^{h} w_i z_i\right)$$

$$z_i = S_h\left(w_{0i} + \sum_{i=1}^{n} w_{ji} x_j\right) \tag{4.33}$$

where $S_o(\ )$ is the output function of the output node in the network which can be sigmoid or linear, $S_h(\ )$ is the output function of each hidden node which is usually sigmoid, w_0 is the bias term of the network output node, w_i is the connection weight between the ith hidden node and the output node, z_i is the output value of the ith hidden node, h is the number of the hidden nodes, w_{0i} is the bias term of the ith hidden node, w_{ji} is the connection weight between the jth input node and the ith hidden node, and n is the number of the network inputs. This equation represents a static relationship between inputs and outputs. It can be applied to the dynamic case as long as the network output $\hat{y}$ and inputs $x_1, x_2,..., x_n$ are replaced with $\hat{y}(t)$ and $y(t-1), y(t-2),..., y(t-p), u(t-1), u(t-2),..., u(t-q)$, respectively. So, the number of network inputs $n = p + q$.

Based on the forward pass equations, the training algorithm of the network can be derived by solving the optimization problem to minimize the following objective function

$$J = \frac{1}{2}\sum_{i=1}^{N}(y_i - \hat{y}_i)^2 \tag{4.34}$$

A variety of methods to solve the optimization problem exist. Rumelhart et al. (1986a) used the method of gradient descent to develop the BP training algorithm for the network. In the BP training algorithm, the weights are updated as follows

$$W^{\text{new}} = W^{\text{old}} + \Delta W^{\text{old}}$$

$$\Delta W^{\text{old}} = \eta \frac{\partial J}{\partial W}\bigg|_{w=w^{\text{old}}} \tag{4.35}$$

where η is a constant as the adaptation rate of network training and the capitalized W represents the set of all weights, small w. This equation indicates that the weights are updated in proportion to the gradient of J with respect to the weights. Rumelhart et al. (1986a) gave the derivation of the BP algorithm. Earlier, Werbos (1974) gave a different version of the derivation. They used different approaches, but the results were exactly the same. In the derivation, Rumelhart used the chain rule directly while in Werbos' derivation, the chain rule was used in a convenient way by ordered derivatives. Werbos' derivation is more general and mathematically rigorous than Rumelhart's.

The details of the derivations can be found in their works. Huang (1995) gave the derivations of the BP algorithm following the ideas of Rumelhart's and Werbos for modeling and control of a snack food frying process. The final equations of the BP algorithm for weight updating based on Eqs. (4.34) and (4.35) are as follows

$$w_i^{new} = w_i^{old} + \eta \delta z_i$$
$$w_0^{new} = w_0^{old} + \eta \delta$$

$$\delta = (y - \hat{y})S_o'\left(w_0^{old} + \sum_{i=1}^{h} w_i^{old} z_i\right) \tag{4.36}$$

$$w_{ji}^{new} = w_{ji}^{old} + \eta \delta_i x_j$$
$$w_{0i}^{new} = w_{0i}^{old} + \eta \delta_i$$

$$\delta_i = \delta w_i^{old} S_h'\left(w_{0i}^{old} + \sum_{j=1}^{n} w_{ji}^{old} x_j\right)$$

where δ and δ_i are the values from the network output node and the ith hidden node which are used to propagate the error at the output back to the input.

With the structure and the training algorithm for modeling a static relationship similar to Eq. (4.3), the input vector, $\underline{x}$, and the output variable, y, can be used to approximate the functional relationship as

$$\hat{y} = \hat{f}(\underline{x}, W) \tag{4.37}$$

where $\underline{x} = (x_1, x_2, ..., x_n)^T$, and W is a matrix representing the connection weights in the network including w_i, w_0, w_{ji}, and w_{0i}. For modeling a dynamic process similar to Eq. (4.5), the input variable, u, and the output variable, y, can be used to approximate the functional relationship as

$$\hat{y}(t) = \hat{f}(y(t-1), y(t-2), ..., y(t-p), u(t-1), u(t-2), ..., u(t-q), W) \tag{4.38}$$

When applying the BP algorithm, the network can be unstable when the learning rate, η, is set too large, which is sometimes done in order to speed up the convergence of the training process. A simple method to increase the learning rate and avoid the instability is to include a momentum term in the update equations of weights (Rumelhart et al., 1986b)

$$W^{new} = W^{old} + \Delta W^{old}$$

$$\Delta W^{old} = \eta \frac{\partial J}{\partial W}\bigg|_{w=w^{old}} + \rho \Delta W_l^{old} \tag{4.39}$$

where ΔW_I^{old} represents the weight update at the last iteration, and ρ is the momentum constant between 0 and 1, which acts on the momentum term to keep the direction of the gradient descent from changing too rapidly when the learning rate is increased.

In addition to the gradient descent method, a number of other optimization methods can be used to increase the efficiency of the BP training process. Conjugate gradient and Levenberg–Marquardt methods have been used in many ANN applications. Choosing a method for nonlinear optimization depends on the characteristics of the problem to be solved. For objective functions with continuous second derivatives (which would include feedforward nets with the most popular differentiable activation functions and error functions), three general types of algorithms have been found to be effective for most practical purposes (Sarle, 1999).

- For a small number of weights, stabilized Newton and Gauss–Newton algorithms, including various Levenberg–Marquardt and trust-region algorithms, are efficient.
- For a moderate number of weights, various quasi-Newton algorithms are efficient.
- For a large number of weights, various conjugate-gradient algorithms are efficient.

The Levenberg–Marquardt algorithm is a standard nonlinear least square technique that can be incorporated into the BP process to increase the training efficiency although the computational requirements are much higher (Hagan and Menhaj, 1994). In order to extend the BP algorithm with the Levenberg–Marquardt method, the objective function of training in Eq. (4.34) can be re-expressed according to Eq. (4.37) as

$$J(W) = \frac{1}{2}\sum_{i=1}^{N}(y_i - \hat{y}_i)^2$$

$$= \frac{1}{2}\sum_{i=1}^{N}[y_i - \hat{f}_i(x, W)]^2$$

$$= \frac{1}{2}\sum_{i=1}^{N}e_i^2(x, W)$$

$$= \frac{1}{2}e^T(x, W) \cdot e(x, W) \tag{4.40}$$

where $e_i^2(x, W) = y_i - \hat{f}_i(x, W)$, and $e(x, W) = [e_1(x, W), e_2(x, W),..., e_N(x, W)]^T$. So, for the following optimization problem,

$$\min_{w^*} J(W) \tag{4.41}$$

where W^* is the optimal matrix of W for the objective function J, using the Gauss–Newton method (Scales, 1985). The following updating equation can be obtained

$$W^{new} = W^{old} - [D^T\underline{e}(\underline{x}, W^{old}) \cdot D\underline{e}(\underline{x}, W^{old})]^{-1} \cdot D^T\underline{e}(\underline{x}, W^{old}) \cdot \underline{e}(\underline{x}, W^{old})$$

$$(4.42)$$

where $D\underline{e}(\underline{x}, W)$ is the first derivative of the vector function $\underline{e}(\underline{x}, W)$ at W, that is, the Jacobian matrix of $\underline{e}(\underline{x}, W)$ at W. In general, W^{new} is a better approximation of W^{old}.

The Levenberg–Marquardt algorithm modified the Gauss–Newton algorithm to the following

$$W^{new} = W^{old} - [D^T\underline{e}(\underline{x}, W^{old}) \cdot D\underline{e}(\underline{x}, W^{old}) + \mu I]^{-1} \cdot D^T\underline{e}(\underline{x}, W^{old}) \cdot \underline{e}(\underline{x}, W^{old})$$

$$(4.43)$$

where I is a unit matrix. The factor μ is important. When it is zero, Eq. (4.43) reverts back to the Gauss–Newton algorithm. When it is large, it becomes the gradient descent algorithm. The values of μ for the Levenberg–Marquardt algorithm occur in between. This algorithm can determine the search direction even when the matrix $D^T\underline{e}(\underline{x}, W) \cdot D\underline{e}(\underline{x}, W)$ is singular. The factor μ is also a convenient parameter to adjust the convergence of the iterative process. Hagan and Menhaj (1994) used this algorithm in neural network training for five approximation problems with significant improvement on the computational efficiency. Huang et al. (1998) investigated the application of a Levenberg–Marquardt algorithm-based BP training algorithm to improve the efficiency of the training processes and generalization of the MFNs in prediction modeling for meat quality evaluation with wavelet textural features from ultrasonic elastograms.

In the BP training process, the output of the feedforward network tends to approximate the target value given the inputs inside the training data set. For the purpose of modeling for prediction, a neural network needs to generalize what is established in training, that is, to let the network output approximate its target value outside the training data set for the given inputs. Several techniques can be used to improve the generalization of a neural network (Finnoff et al., 1993). Among them, weight decay is effective. Weight decay adds a penalty term to the regular objective function for network training

$$J_{wd} = J + \lambda J_{pt} \qquad (4.44)$$

where J_{wd} is the overall objective function for weight decay, J_{pt} is the penalty term that causes the weights to converge to smaller absolute values than they would in the regular case, and λ is the decay constant, which can be tuned for the performance of the weight decay.

For the minimization problem of J_{wd}, the optimal vector W^* should satisfy the following equation

$$\nabla J_{wd} = 0 \qquad (4.45)$$

where ∇ represents the gradient operator. If the Gauss–Newton method is used, and the penalty term is the squared sum of all weights in the network, the weight updating equation is

$$W = [D^T \underline{e}(\underline{x}, W) D\underline{e}(\underline{x}, W) + 2\lambda WI]^{-1} \cdot [D^T \underline{e}(\underline{x}, W) D\underline{e}(\underline{x}, W)] W$$
$$- [D^T \underline{e}(\underline{x}, W) D\underline{e}(\underline{x}, W) + 2\lambda WI]^{-1} \cdot D^T \underline{e}(\underline{x}, W) \underline{e}(\underline{x}, W) \qquad (4.46)$$

The generalization ability of a network depends on the decay constant. With weight-decay training, the network can avoid oscillation in the outputs caused by large weights.

The leave-one-out procedure (Duda and Hart, 1973) can be used to validate only one sample at a time and train the network with the rest of the samples. This can be useful where the total sample size is small. The procedure used in network training is as follows:

1. Set the counter $c = 1$.
2. For the data set with N samples, set up a training data subset containing $N - 1$ input–output pairs excluding the cth pair.
3. Train the network with the training data until the training process converges satisfactorily.
4. Validate the network using the excluded input–output pair.
5. If $c = N$, then finish the training process. Otherwise, let $c = c + 1$ and go to Step 2.

The generalization ability of the network can be assessed by evaluating all of the validations done in the implementation of the procedure.

The BP algorithm is an example of supervised training algorithms. Supervised training algorithms use an external teacher or input–target (output) pairs. In supervised training, inputs are applied to the network and the network's output is calculated and compared to the target values. The difference or error is propagated back to the network. With the back propagated error, the network weights are adjusted in a manner to reduce the error in the next iteration. Unlike supervised training, unsupervised training has no "teacher," and input patterns are applied to the network and the network self-organizes by adjusting the weights according to a well-defined algorithm and network structure.

One important self-organizing principle of sensory pathways in the brain is that the position of neurons is orderly and often reflects some physical characteristics of the external stimulus being sensed. Kohonen presented such a self-organizing network that produces what he called self-organizing feature maps (SOM) similar to those that occur in the brain (Kohonen, 1984; Lippman, 1987). Kohonen's self-organizing network is an important method in unsupervised training.

Kohonen's self-organizing network is a two-layer structure. The network's output nodes are arranged orderly in a one- or two-dimensional array, and every input is connected to every output node by an adjustable connection weight.

There are two steps in the training process of a Kohonen self-organizing network. The first step is to generate a coarse mapping. In this step, the neighbors, NE_{j^*} as an example for the one-dimensional case, are defined for updating as

$$NE_{j^*} = (\max(1, j^* - 1), j^*, \min(m, j^* + 1)) \qquad (4.47)$$

This means that every output node j has the neighbors $j - 1$ and $j + 1$, except at the borders of the array, whereby the neighbor of node 1 is 2, and the neighbor of the node m is $m - 1$, respectively. Also in this step, in order to allow for large weight modifications and to settle into an approximate mapping as quickly as possible, the adaptation gain term, η, should remain high (>0.5).

The second step is to fine-tune the output nodes to the input vectors within localized regions of the map. In this step, the neighbors are reduced to winning nodes only, that is, $NE_{j^*} = (j^*)$, and the adaptation gain term, η, should be low (<<0.5).

The Kohonen self-organizing network has been successfully applied in areas such as speech recognition (Aleksander, 1989) and image processing (Nasrabadi and Feng,1988). Whittaker et al. (1991) used it for ultrasonic signal classification in beef quality grading.

In an ANN model, the final model needs not only to perform satisfactorily for the training data, but also to produce acceptable outputs when presented with input data not used in the training data. In general, the neural network training error is different than the error generated in validation of the neural network. The neural network process model with acceptable training error and generalization should have the smallest structure, that is, the minimum number of hidden nodes and past process inputs and outputs.

Usually two sets of data are grouped from the process input–output data without overlap. One set of data is used to train the network, and the other is used to test or validate the network model. In general, with the increase of model complexity, that is, the increase of model order or the number of network hidden nodes, the model error measurement on the training data decreases continuously and the model error measurement on the test or validation data decreases initially, then reaches a minimum point, and increases again afterwards. This behavior can be used to determine the structure parameters of a neural network with the cross validation of the two data sets. As the number of model structure parameters, model order, or number of network hidden nodes increases, the model training error decreases significantly. Although the training error still decreases, this decrease is not significant, and the model generalization is not improved after the minimum point of the test or validation error. Too many model structure parameters may result in a network model that overfits the data. Obviously, the model with the complexity corresponding to the minimum point of the test or validation error is the best choice. The number of model structure parameters corresponding to this point determines the smallest structure of the model.

The model error measurement on the training data and the number of model structure parameters can be combined to make statistical criteria that are used to help find the smallest model structure. Model identification is a process of model selection in which a number of models are fitted, and then the best one is determined. Shibata (1985) surveyed a variety of criteria for model selection. It has been found that during the process of model selection, model overfitting and underfitting can be balanced effectively if a given criterion contains not only a model training error measurement but also an assignment of cost to the introduction of each additional parameter in the model. A number of criteria based on this idea are in this form

$$C = F(TE, N_0) \qquad (4.48)$$

where TE is any statistic which measures the network model training error, N_0 is the total number of parameters in the model, and C or $F(\ )$ represents the quantity of a criterion. The criterion is formed to penalize the models with a large N_0. A large TE rejects overly simple models, and a high penalty term rejects highly complex models. Therefore, a model that achieves the least C is the final product in modeling. Earlier than AIC described in Section 4.2 of this chapter, Akaike (1969) proposed such a criterion named Akaike's final prediction error (FPE)

$$FPE = \frac{N + N_0}{N - N_0} TSE \qquad (4.49)$$

where TSE is the summed squared error of the training data. The FPE rejects overly simple models and penalizes highly complex models. The final model is determined when the FPE reaches the minimum point. The penalty term of the FPE works gradually in terms of the given nonlinear function and the TSE multiplies the penalty term to constitute the criterion.

Sufficient numbers of hidden layers and hidden nodes are essential for a neural network to approximate a nonlinear function. The order of a model is important for a neural network to fit and generalize dynamic process input–output data accurately. One hidden layer in a neural network provides sufficient function approximation ability (Hornik et al., 1989; Cybenko, 1989). The determination of the number of hidden nodes in a neural network along with the determination of the order of the network model is crucial in the identification of a neural network process model. A cross-validation procedure can be used to determine the number of network hidden nodes and model order. As illustrated previously, the training error decreases as the number of hidden nodes or model order increases. However, as the number of network hidden nodes or model order increases, the test or validation error reaches a minimum point and then increases. The FPE may have a minimum value of the network test or validation error, so it can be used to cross validate the network training and test or validation errors.

Huang et al. (1998a) gave a specific cross-validation procedure for identification of the smallest structure of the neural network process models as follows

1. Start the network hidden nodes or model order with 1.
2. Set a seed for the random number generator.
3. Initialize all of the weights to small random numbers.
4. Train the network.
5. Compute the FPE of the network process model.
6. Validate the network.
7. Check whether the validation error or the FPE reaches a minimum point. If any minimum point is reached, stop. If the minimum point is not yet reached, make an increment of 1 for hidden node or model order and repeat from Step 3.

In order to have a common starting point for training different-sized networks, all weights in each updated network are initialized to small random numbers using the random number generator with a fixed seed. In this way, the overlapped weights remain the same although for each updating the number of weights is different.

The FPE is incorporated in this procedure to verify the minimum point alternately with the validation error. In general, the minimum point of the FPE may be consistent with the minimum point of the validation error. However, in some cases, they are inconsistent. When this happens, the network models corresponding to the two different minimum points will be compared to make the final decision.

4.3.1 Example: ANN modeling for beef grading

Whittaker et al. (1991) used two ANN approaches (supervised and unsupervised training) and a conventional statistical approach to develop classifiers for beef grading based on ultrasonic signals. The classifiers were trained for all of the approaches using the same data set containing 100 individual meat samples. These samples covered the marbling ranges typically found in U.S. markets. In addition, the majority of the samples fell into the most important economic quality grades—USDA Select and Choice. The classifiers were evaluated using a separate data set of 24 samples. The validation data set had a very similar distribution to the training set.

Only the 2.25 MHz probe data was used for supervised training. Convergence of the BP training process was not obtained with all seven frequency parameters because *a priori* information from statistical analysis was used to select fewer input variables, that is, f_c, f_{sk}, and Lm. The performance of these classifiers was evaluated by the accuracy of classification, which will be discussed in the corresponding example in the next chapter.

In order to evaluate the optimal effectiveness of the network in the unsupervised training, experiments were performed according to the number of inputs (2 to 7 ultrasonic frequency parameters) and the number of output nodes (2 to 8 classes). There existed 119 ($C_7^2 + C_7^3 + C_7^4 + C_7^5 + C_7^6 + C_7^7$) possibilities of the combinations of input variables, so there were 833 (7 × 119) experiments. Obviously, conducting so many experiments was not worthwhile. Therefore, through experimentation and *a priori* statistical knowledge, important combinations were selected.

In the experiments, different combinations of 7 ultrasonic A-mode signal features were used and classified into 3, 4, and 8 classes, which are also presented in the next chapter.

According to Kohonen's recommendation, the number of training cycles should be at least 500× the number of output nodes for good statistical accuracy. For a different number of classes, that is, the number of output nodes, the required numbers of training cycles were different. For instance, for 3 output nodes, the required number of training cycles was 1500 (3 × 500); for 4 output nodes, the number was 2000 (4 × 500); and for 8 output nodes, the number was 4000 (8 × 500). In training the adaptation gain term, η, was taken in the expression of an exponent function of the training step, k:

$$\eta(k) = 0.99e^{-k/T}$$

When $k = 0$, $\eta(0) = 0.99$ which is very close to 1. In order to reasonably divide the training process into two steps (coarse mapping and fine-tuning), the required number of training cycles, T, should be different. For instance, for 3 output nodes, T should be 200; for 4 output nodes, T should be 250; and for 8 output nodes, T should be 2000.

4.3.2   Example: ANN modeling for food odor pattern recognition by an electronic nose

Because many of the responses are nonlinear with respect to odorant concentration, ANNs are usually the most successful at coding the sensor response. Typically, ANNs model the nonlinear relationship between sensor readings and sample classification assignments

$$\hat{c} = \hat{f}(\tilde{x}_1, \tilde{x}_2, ..., \tilde{x}_n; \hat{\Theta}) \tag{4.50}$$

where $\hat{f}(\,)$ is the nonlinear function estimate, $\hat{\Theta}$ is the set of coefficient estimates, and other symbols are the same as in Eq. (4.29).

However, a significant amount of work on neural network methods for application to electronic noses is needed. Applications of electronic noses thus far have involved a supervised training process. The electronic nose must be trained by presenting several examples of known odors. The target values of these odors, whether quantitative, as with intensity, or qualitative, as with odor hedonics, must be known independently of the electronic nose. Often this means that olfactometry data must be collected on the training samples in order to establish the target values for the neural network. When presented with an odor on which no training data exists, the electronic nose is unable to classify the sample from the catalog of known odors. In this aspect, the electronic nose is similar to human response which is shaped through experience.

A variety of ANN architectures have been applied to specific problems with the electronic nose. In many cases, these strategies include the application of fuzzy membership functions to enhance classification. Determining the appropriate ANN for a given problem is currently an empirical process and, consequently, time consuming. Once established, however, the selected ANN model can be used to rapidly classify a new series of data for a particular odor.

4.3.3   Example: ANN modeling for snack food eating quality evaluation

A statistical analysis was performed to select prominent image features that define the quality in terms of the sensory attributes. Then these prominent texture and morphology features were determined as input to a model for quality classification. The purpose of this analysis was to reduce the model size and, hence, the computing time. In this analysis, textural, size, and shape features were considered to be independent variables, and the sensory attributes were dependent variables.

Through two separate stepwise regressions, 11 features (5 textural features, F_1, F_3, F_6, F_9, and F_{10}, and 6 size–shape features, PERIM, FIBERW, LENGTH, ROUND, ASPR and FULLR) were selected as the input to the classification model, that is, the classifier. When all 22 features were regressed together, 8 textural and shape features, F_5, F_{12}, PERIM, LENGTH, FIBERW, ROUND, FULLR, and ASPR, were determined to be the prominent ones. The originally considered textural features (F_1, F_3, F_6, F_9, and F_{10}) were excluded because of their high correlation with the shape features, and the textural features (F_5, F_{12}) were included because of their low correlation with the shape features.

A one-hidden-layer ANN trained with BP was used to model the relationship between textural and morphological features of snack images and snack sensory attributes.

Consider a set of N vector pairs, (x_1, y_1), (x_2, y_2),...,(x_N, y_N), where x and y are related by some nonlinearity. In our case, x represents textural and morphological features and y represents sensory attributes. The neural network is trained to learn the unknown nonlinear relationship. This is achieved by minimizing the mean square error (MSE) between the desired and actual sensory attributes. The size of the network input layer is equal to the size of the feature vector, that is, 22. The size of the output layer is taken to be the number of the output attributes, that is, seven. However, the size of the hidden layer should be obtained by experiments. In our case, the number

Table 4.5 Training and Validation Samples for Different
Experimental Setups*

Machine Wear–Raw Material Conditions	Number of Training Samples	Number of Validation Samples
A	700	100
B	400	50
C	400	50
D	400	100

* Adapted from Sayeed et al. (1995). With permission.

of hidden nodes was incremented by 1 starting from one hidden node until the mean square error was reached at 0.1 percent. Hence, the smallest amount of hidden nodes, leading to the smallest amount of training time that achieved convergence was nine. Thus a $22 \times 9 \times 7$ network was structured. The network outputs were normalized to real numbers between 0 and 1. The taste panel grading from –3 to 3 was also mapped to a value between 0 and 1.

The machine conditions play an important role in the formation of the snacks. The raw material conditions are also important in the formation of the snacks. With the preset conditions, the samples were BP trained by neural networks consisting of 22 input nodes, 9 hidden nodes, and 7 output nodes designed to model the nonlinear relationship between the input textural–morphological features divided into four "machine wear–raw material categories (A, B, C, and D)." A total of 50 samples per cell constituted the training and validation samples for ANN classification. Table 4.5 shows the number of randomly arranged training and validation samples used.

4.3.4 Example: ANN modeling for meat attribute prediction

Data from elastography analysis and mechanical and chemical tests for beef samples in LD muscle were available for modeling to predict the beef quality attributes. Huang et al. (1998) used the data to build the neural network prediction models with the Levenberg–Marquardt's algorithm for effective training and network generalization. There were 29 sample vectors in the data. Each vector included the wavelet image textural feature parameters from each beef sample from the elastograms as the inputs and the mechanical and chemical measurements as outputs. Because each of the elastograms was cropped into a 128×128 image, the number of feature inputs was 29. The mechanical and chemical measurements were used as the indicators of beef tenderness.

The modeling work was implemented using the regular BP algorithm for training with and without a momentum term, and then using the Levenberg–Marquardt BP algorithm for training with and without weight decay. The leave-one-out procedure was built in all of the training processes. The performance difference of these training processes was then compared. The programs were coded using MATLAB (The MathWorks, Inc.).

Table 4.6 Result of the Determination of the
Number of Hidden Nodes for the
Network Model of WBSF*

Number of Hidden Nodes	Validation MSE
1	0.1294
2	0.0410
3	0.0414
4	0.0416

* From Huang et al. (1998). With permission.

Table 4.7 Results of the Determination of the Number of Hidden Nodes for the
Models with Daubechies-4 Wavelet Features in the Regular Leave-One-Out
BP Training without a Momentum Term*

Model Output	No. of Hidden Nodes	Training R^2	Validation MSE	No. of Training Epoches	No. of Flops
WBSF	2	0.95	0.0207	30.48×10^4	25.48×10^8
Calp	3	0.96	0.0203	44.58×10^4	54.85×10^8
Sarc	2	0.91	0.0387	17.40×10^6	14.54×10^9
T. Coll	4	0.96	0.0207	29.08×10^4	47.25×10^8
%Sol	4	0.95	0.0218	47.27×10^3	76.81×10^7
%Mois	2	0.95	0.0214	10.97×10^5	91.70×10^8
%Fat	4	0.95	0.0205	98.89×10^2	16.07×10^7

* From Huang et al. (1998). With permission.

As the first step, the model structure of the neural networks, that is, the number of hidden nodes for the one-hidden-layered network, was identified. The method of identification was to determine the optimal number of hidden nodes by selecting the number with the lowest validation MSE value of a network model by the increment of the number of the hidden nodes from 1, 2,..., H where H was a number, 4 or 6, as long as an optimal number with the lowest validation MSE was found. Table 4.6 shows the MSE for 1 to 4 hidden nodes for the neural network model of WBSF, where 2 nodes was the optimal number. Table 4.7 shows the results of the determination of the number of hidden nodes for the models based on inputs of Daubechies-4 wavelet features using the leave-one-out regular BP training without a momentum term. In order to present the efficiency of the training processes completely, both the number of training epochs and the number of floating point operations (flops) for the training computation were recorded. The learning rate was $\eta = 0.025$, and the training stopping criterion was MSE = 0.01 in order to ensure the stability of the training processes. If it is desirable for the training process to be accelerated, the learning rate should be increased. However, if the learning rate parameter η is increased arbitrarily for the purpose of increasing the learning rate, the network may become unstable. To overcome this problem, the learning rate of the training

Table 4.8 Results of Convergence Acceleration Using a Momentum Term
in the Regular Leave-One-Out BP Training with
Daubechies-4 Wavelet Features*

Model Output	Training R^2	Validation MSE	ρ	No. of Training Epochs	No. of Flops
WBSF	0.95	0.0207	0.90	29.61×10^4	25.13×10^8
Calp	0.96	0.0200	0.60	44.45×10^4	55.55×10^8
Sarc	0.91	0.0387	0.025	17.40×10^6	14.77×10^9
T. Coll	0.96	0.0207	0.90	10.57×10^4	17.45×10^8
%Sol	0.96	0.0204	0.05	29.46×10^3	48.63×10^7
%Mois	0.95	0.0206	0.90	13.35×10^4	11.33×10^8
%Fat	0.95	0.0205	0.025	98.89×10^2	16.32×10^7

* From Huang et al. (1998). With permission.

Table 4.9 Ratios of Training Epochs and Flops without and with
a Momentum Term in the Regular Leave-One-Out BP
Training Using Daubechies-4 Wavelet Features*

Model Output	Ratio of Epochs	Ratio of Flops
WBSF	1.03	1.01
Calp	1.00	0.99
Sarc	1.00	0.98
T. Coll	2.75	2.71
%Sol	1.60	1.58
%Mois	8.22	8.09
%Fat	1.00	0.98

* From Huang et al. (1998). With permission.

process can be increased and the network kept stable at the same time by introducing a momentum term in weight updating, that is, let $0 < \eta < 1$. Table 4.8 shows the results of leave-one-out BP training with a momentum term. Table 4.9 shows the ratios of training epochs and the number of flops with and without momentum terms. Training with momentum terms took more flops for the models of Sarc and %Fat where the number of training epochs remained the same. Training with momentum terms for the model of Calp also increased flops but the number of training epochs was reduced a bit. The training with momentum term for the rest of the models took a range of flops reduced by a factor of 0.01 up to 7.09× where the training epochs were reduced.

Incorporating the Levenberg–Marquardt algorithm into the leave-one-out BP training is another possibility to accelerate the training process. Table 4.10 shows the results of Levenberg–Marquardt leave-one-out BP training. The factor μ was determined as 1 in all the cases in order to ensure that the algorithm converged effectively. All of the models converged to the given error criterion with much less training epochs. Table 4.11 shows the ratios of the training epochs and number of flops before and after using the

Table 4.10 Results of the Leave-One-Out BP Training Using the
Levenberg–Marquardt Algorithm with the
Daubechies-4 Wavelet Features*

Model Output	Training R^2	Validation MSE	μ	No. of Training Epoches	No. of Flops
WBSF	0.97	0.0155	1.00	580	20.90×10^7
Calp	0.98	0.0120	1.00	464	44.06×10^7
Sarc	0.95	0.0296	1.00	1827	72.13×10^7
T. Coll	0.96	0.0200	1.00	493	89.98×10^7
%Sol	0.97	0.0158	1.00	464	83.07×10^7
%Mois	0.96	0.0159	1.00	580	22.44×10^7
%Fat	0.96	0.0193	1.00	261	47.40×10^7

* From Huang et al. (1998). With permission.

Table 4.11 Ratios of Training Epochs and Flops before and after Using
the Levenberg–Marquardt Algorithm in the Leave-One-Out BP
Training with Daubechies-4 Wavelet Features*

Model Output	Ratio of Epochs	Ratio of Flops
WBSF	525.60	12.19
Calp	960.75	12.45
Sarc	9523.81	20.16
T. Coll	589.82	5.25
%Sol	101.88	0.92
%Mois	1891.75	40.87
%Fat	37.89	0.34

* From Huang et al. (1998). With permission.

Levenberg–Marquardt algorithm in the leave-one-out BP processes. From Table 4.10, it can be seen that all models obtained higher R^2 values and lower validation MSE values. This indicates that these network models have better output variation accounting and generalization. This may be related to the flexibility in the convergence space of the Levenberg–Marquardt algorithm. Table 4.11 further indicates that after using the Levenberg–Marquardt algorithm for this application, the number of training epochs was reduced greatly. However, when the reduction was only 100× or less, the number of flops was more, such as the models of %Sol and %Fat. This means that in each iteration step, the Levenberg–Marquardt algorithm needed more operations. Thus, this algorithm had a higher computation requirement. In the cases where the epoch reduction was over several hundred times, the number of flops required for training was reduced approximately from 4 to 40×.

Further, on the basis of the use of the Levenberg–Marquardt algorithm, the weight-decay algorithm can be incorporated to improve the network generalization. Basically, the weight-decay algorithm, as described previously, suppressed excessively large weights in the network to maintain the

Table 4.12 Results of the Leave-One-Out BP Training Using Weight-Decay with the Levenberg–Marquardt Algorithm in Daubechies-4 Wavelet Features*

Model Output	Training R^2	Validation MSE	λ	No. of Training Epoches	No. of Flops
WBSF	0.98	0.0097	0.01	522	19.52×10^7
Calp	0.99	0.0044	0.001	406	34.45×10^7
Sarc	0.98	0.0086	0.01	464	16.29×10^7
T. Coll	0.98	0.0079	0.01	493	87.86×10^7
%Sol	0.99	0.0052	0.01	435	82.51×10^7
%Mois	0.98	0.0054	0.01	493	18.83×10^7
%Fat	0.98	0.0068	0.01	261	47.40×10^7

* From Huang et al. (1998). With permission.

network stability and to reduce noise fitting in modeling. Table 4.12 shows that incorporating weight-decay into the leave-one-out Levenberg–Marquardt BP training achieved better models which had higher R^2 and lower validation MSE values. The decay constant λ was determined to see if it gave a better model generalization. The results indicated that these models had better output variation accounting and generalization. These models were evaluated as the best models with good output variation accounting and less noise fitting. It was also interesting that incorporating weight decay made the training slightly more efficient vs. implementing the Levenberg–Marquardt algorithm alone.

4.3.5 Example: ANN modeling for snack food frying process control

As described earlier, there are several factors affecting the product quality of the continuous, snack food frying process. Huang et al. (1998a) studied the process based on the structure of a 2×2 system in modeling for the purpose of process control. In the 2×2 system, the inlet temperature of the frying oil and the residence time of the product in the fryer were identified as significant control variables that affect the final product quality. Sensors were placed at the end of the production line to measure the product quality attributes of color and moisture content, which are controlled variables, to indicate the final product quality. This 2×2 process is governed by the following discrete-time, time-delayed, NARX system equation

$$y(t) = f(y(t-1), y(t-2),\dots, y(t-p), u(t-d-1),$$

$$u(t-d-2),\dots, u(t-d-q), \Theta) + \varepsilon(t) \tag{4.51}$$

where $y(t) = [y_1(t), y_2(t)]^T$. It is the process output vector in which $y_1(t)$ is color and $y_2(t)$ is the moisture content (percent) at time t; $u(t) = [u_1(t), u_2(t)]^T$. It is the process input vector in which $u_1(t)$ is inlet temperature (°C) and $u_2(t)$ is residence time (5 s) at time t: $\Theta = (\Theta_1, \Theta_2)^T$ is the set of parameters in which Θ_1

is for the equation of $y_1(t)$ and Θ_2 is for the equation of $y_2(t)$: d represents the time lags between the process input $\underline{u}(t)$ and the process output $\underline{y}(t)$, that is, $\underline{u}(t-d) = [u_1(t-d_1), u_2(t-d_2)]^T$ in which d_1 and d_2 are the minimum time lags of the inputs related to the present outputs $y_1(t)$ and $y_2(t)$, respectively: $\underline{\varepsilon}(t) = [\varepsilon_1(t), \varepsilon_2(t)]^T$ is the measurement noise vector in which $\varepsilon_1(t)$ is for the equation of $y_1(t)$ and $\varepsilon_2(t)$ is for the equation of $y_2(t)$ at time t, which is assumed to be two-dimensional Gaussian distributed with zero mean and constant variance.

Eq. (4.4) provided a more general description of this representation. In Eq. (4.47), the time lags are represented explicitly with the symbol d (5 s) because in the process any action from the actuators for controlling the temperature and the conveyor speed will take a period of time to take effect on the sensor readings for color and moisture content. These time lags were determined through field measurement and statistical analysis. They were 20 and 16 units (5 s), respectively, from inlet temperature, residence time to the process outputs, color and moisture content.

ANNs could be used to directly approximate the function $f(\ )$ as described in Eq. (4.51) as follows

$$\hat{\underline{y}}(t) = \hat{f}(\underline{y}(t-1), \underline{y}(t-2),\ldots, \underline{y}(t-p), \underline{u}(t-d-1),$$
$$\underline{u}(t-d-2),\ldots, \underline{u}(t-d-q), W) \tag{4.52}$$

This modeling approach has no network output feedback and takes the structure of a MFN. It can provide the one-step-ahead predictor, described in the next chapter, for internal model control (IMC) which is described in Chapter 6.

A one-hidden-layered feedforward neural network was trained with the BP algorithm to model the 2×2 system. Before training, the data were scaled as Eq. (3.1). Following the procedure of model identification for neural networks, the smallest structure of the neural network process model was identified. From Table 4.13, it can be seen that although the FPE still decreased, 3 was chosen as the number of hidden nodes of the neural network process model because the test MSE had a minimum there. Table 4.14 shows that the test MSE and FPE had the minimum values at (2, 2, 2, 2) of the

Table 4.13 Results of the Determination of the Number of Hidden Nodes of the Neural Network Model for the Snack Food Frying Process*

Number of Hidden Nodes	Training MSE	Test MSE	FPE
1	0.107641	0.132214	0.108852
2	0.031462	0.051293	0.032226
3	0.029928	0.050594	0.030994
4	0.029437	0.050692	0.030823
5	0.028460	0.052278	0.030130

* From Huang et al. (1998a). With permission.

Table 4.14 Results of the Order Determination of the Neural Network
Model for the Snack Food Frying Process*

Model Order	Training MSE	Test MSE	FPE
(1, 1, 1, 1)	0.032226	0.055117	0.032648
(2, 2, 2, 2)	0.029928	0.050594	0.030655
(3, 3, 3, 3)	0.031292	0.052863	0.032407
(4, 4, 4, 4)	0.034187	0.054443	0.043841

* From Huang et al. (1998a). With permission.

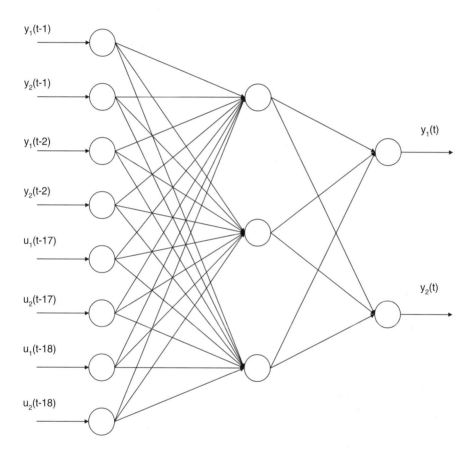

Figure 4.9 Resulting structure of the neural network model for the snack food frying
process.

model order (p_1, p_2, q_1, q_2). The order of the neural network model was
determined as (2, 2, 2, 2). The resulting smallest structure of the neural
network model was $8 \times 3 \times 2$, that is, 8 inputs by 3 hidden nodes by 2 outputs,
which is shown in Figure 4.9.

References

Akaike, H., Fitting autoregressive models for prediction, *Ann. Inst. Stat. Math.*, 21, 243, 1969.

Akaike, H., Information theory and an extension of the maximum likelihood principle, *Proc. 2nd Int. Symp. Inf. Theor.*, Budapest, 1973, 267.

Aleksander, I., *Neural Computing Architecture*, MIT Press, Cambridge, MA, 1989.

Bullock, D., Modeling of a continuous food process with neural networks, M.Sc. thesis. Texas A&M University, College Station, TX, 1995.

Cybenko, G., Approximator by superpositions of a sigmoidal function, *Math. Contr. Sign. Syst.*, 2, 303, 1989.

Duda, R. O. and Hart, P. E., *Pattern Classification and Scene Analysis*, John Wiley & Sons, New York, 1973.

Efroymson, M. A., Multiple regression analysis, in *Mathematical Methods for Digital Computers*, Ralston, A., and Wilf, H. S., Eds., John Wiley & Sons, New York, 1962.

Finnoff, W., Hergert, F., and Zimmermann, H. G., Improving model selection by nonconvergent methods, *Neur. Net.*, 6, 771, 1993.

Grossberg, S., Adaptive pattern classification and universal recording. I. Parallel development and coding of neural feature detectors, *Biol. Cybern.*, 23, 121, 1976.

Hagan, M. T. and Menhaj, M. B., Training feedforward networks with the Marquardt algorithm, *IEEE Trans. Neur. Net.*, 596, 989, 1994.

Haralick, R. M., Shanmugam, K., and Dinstein, I., Textural features for image classification, *IEEE Trans. Syst., Man, Cybern.*, 3(6), 610, 1973.

Hebb, D. O., *The Organization of Behavior*, John Wiley & Sons, New York, 1949.

Hopfield, J. J., Neural networks and physical systems with emergent collective computational abilities, *Proc. Nat. Acad. Sci.*, 79, 2554, 1982.

Hornik, K., Stinchcombe, M., and White, H., Multilayer feedforward neural networks are universal approximators, *Neur. Net.*, January, 359, 1989.

Huang, Y., Snack food frying process input–output modeling and control through artificial neural networks, Ph.D. dissertation, Texas A&M University, College Station, TX, 1995.

Huang, Y., Lacey, R. E., Moore, L. L., Miller, R. K., Whittaker, A. D., and Ophir, J., Wavelet textural features from ultrasonic elastograms for meat quality prediction, *Trans. ASAE*, 40(6), 1741, 1997.

Huang, Y., Lacey, R. E., and Whittaker, A. D., Neural network prediction modeling based on elastographic textural features for meat quality evaluation, *Trans. ASAE*, 41(4), 1173, 1998.

Huang, Y., Whittaker, A. D., and Lacey, R. E., Neural network prediction modeling for a continuous, snack food frying process, *Trans. ASAE*, 41(5), 1511, 1998a.

Kohonen, T., *Self-Organization and Associative Memory*, Springer-Verlag, Berlin, 1984.

Lacey, R. E. and Osborn, G. S., Applications of electronic noses in measuring biological systems, ASAE paper No. 98-6116, St. Joseph, MI, 1998.

Lippman, R. P., An introduction to computing with neural nets, *IEEE ASSP Mag.*, April, 4, 1987.

Ljung, L., *System Identification: Theory for the User*, 2nd ed., Prentice-Hall, Upper Saddle River, NJ, 1999.

Lozano, M. S. R., Ultrasonic elastography to evaluate beef and pork quality, Ph.D. dissertation, Texas A&M University, College Station, TX, 1995.

McCulloch, W. S. and Pitts, W., A logical calculus of the ideas immanent in nervous activity, *Bull. Math. Biophys.,* 9, 127, 1943.

Milton, J. S. and Arnold, J. C., *Introduction to Probability and Statistics: Principles and Applications for Engineering and the Computing Sciences,* 2nd ed., McGraw-Hill, New York, 1990.

Minsky, M. L. and Papert, S. A., *Perceptrons,* MIT Press, Cambridge, MA, 1969.

Moore, L. L., Ultrasonic elastography to predict beef tenderness, M.Sc. thesis, Texas A&M University, College Station, TX, 1996.

Moreira, R. G., Palau, J., and Sun, X., Simultaneous heat and mass transfer during deep fat frying of tortilla chips, *J. Food Proc. Eng.,* 18, 307, 1995.

Nasrabadi, N. M. and Feng, Y., Vector quantization of images based upon the Kohonen self-organizing feature maps, *Proc. IEEE Int. Conf. Neur. Net.,* 1, 101, 1988.

Park, B., Non-invasive, objective measurements of intramuscular fat in beef through ultrasonic A-model and frequency analysis, Ph.D. dissertation, Texas A&M University, College Station, TX, 1991.

Rosenblatt, F., The perceptron: a probabilistic model for information storage and organization in the brain, *Psychol. Rev.,* 65, 386, 1958.

Rumelhart, D. E. and McClelland, J. L., *Parallel Distributed Processing: Explorations in the Microstructures of Cognition,* Vols. I & II, MIT Press, Cambridge, Massachusetts, 1986.

Rumelhart, D. E., Hinton, G. E., and Williams, R. J., Learning representations by back-propagating errors, *Nature,* 323, 533, 1986a.

Rumelhart, D. E., Hinton, G. E., and Williams, R. J., Learning internal representation by error propagation, in *Parallel Distributed Processing: Explorations in the Microstructures of Cognition,* Vol. I, MIT Press, Cambridge, MA, 1986b.

Sarle, W. S., *Neural Nets FAQ,* SAS Institute Inc., Cary, NC, 1999.

Sayeed, M. S., Whittaker, A. D., and Kehtarnavaz, N. D., Snack quality evaluation method based on image features and neural network prediction, *Trans. ASAE,* 38(4), 1239, 1995.

Scales, L. E., *Introduction to Nonlinear Optimization,* Springer-Verlag, New York, 1985.

Shibata, R., Various model selection techniques in time series analysis, in *Handbook of Statistics,* Vol. 5, 179, North-Holland, Amsterdam, 1985.

Werbos, P. J., Beyond regression: new tools for prediction and analysis in the behavior sciences, Ph.D. dissertation, Harvard University, MA, 1974.

Whittaker, A. D., Park, B., McCauley, J. D., and Huang, Y., Ultrasonic signal classification for beef quality grading through neural networks, Proc. 1991 Symp. Automat. Agric. 21st century, St. Joseph, MI, 1991.

Whittaker, A. D., Park, B., Thane, B. R., Miller, R. K., and Savell, J. W., Principles of ultrasound and measurement of intramuscular fat, *J. Anim. Sci.,* 70, 942, 1992.

chapter five

Prediction

5.1 Prediction and classification

Prediction is performed by a quantitative target variable through the operation of the model which is typically a regular linearly regressed function. For example, based on ultrasonic scans, mechanical, chemical, and sensory attributes of beef samples can be predicted by a linear or nonlinear regression prediction model. Classification, as opposed to prediction, is performed by a categorical target variable through the model, typically, a discriminant function. For example, based on ultrasonic scans, beef samples can be classified according to USDA's standards by a linear or nonlinear discriminant function.

The purposes of modeling in food quality quantization and process control are sample classification, attribute prediction, and controller design. Sample classification and attribute prediction are usually based on a static mapping of process inputs and outputs, while controller design needs support from process dynamic models that provide one-step-ahead or multiple-step-ahead predictions, depending on the control schemes used in the control loops.

For the problems of sample classification and attribute prediction, refer to Eq. (4.3), at which the input vector $\underline{x}$ and the output vector $\underline{y}$ are assumed to relate to each other with a function $f(\ )$. In the process of modeling, the function parameter vector Θ is estimated and, hence, the function is estimated whether it is linear or nonlinear. Based on the function parameter estimation, the equation for classification or prediction can be represented as

$$\hat{\underline{y}} = \hat{f}(\underline{x}, \hat{\Theta}) \tag{5.1}$$

In this equation, the input vector contains the parameters by the measured data of food samples, and the output vector generates the classification indices or attribute values of the food samples. The input parameters can be the specifically measured data or a transformation of the data in the same or reduced dimension. Reduced dimension is referred to as feature extraction from the input parameter space. The features can be extracted from sensor readings or image quantization to facilitate the computation of modeling for prediction and classification. Each of the classification indices is usually assigned binary numbers; for example, a good apple sample is labeled 1, and bad apple is labeled 0. Attribute values can be from survey or experiments. For example, beef sensory attributes may include values in tenderness, juiciness, flavor, and so on.

As discussed previously, the function $f(\ )$ may be linear or nonlinear. It can be built by linear and nonlinear statistical analysis or ANNs for prediction or classification. In the area of classification, the function $f(\ )$ is usually called a classifier, a term adopted from pattern recognition. In the area of prediction, the function $f(\ )$ is called a predictor, a term adopted from dynamic process modeling.

5.1.1   Example: Sample classification for beef grading based on linear statistical and ANN models

As described in the last chapter, two ANN approaches in supervised and unsupervised training algorithms along with statistical analysis were developed for beef classification in quality grading based on ultrasonic A-mode signals. Table 5.1 presents a summary of the results obtained with BP training. The outputs were processed in a *winner-take-all* fashion, that is, the node with the largest value was declared the winner. The accuracy was calculated simply as the number of correct classifications divided by the total number of samples in the validation data set.

Table 5.2 gives the classification results for the adaptive logic network. Increasing the quantization level increased the accuracy of encoding, but also slightly increased the mean error. Lower levels of quantization also slightly increased the mean error. The number of training pairs varied from 93 to 97. A total of 24 samples were used for classification.

In the experiments of unsupervised training, different combinations of seven ultrasonic A-mode signal features were used and divided into three, four and eight classes as shown in Table 5.3.

Table 5.1 Classification of Marbling Levels with Back Propagation Training*

Probe	Number of Training Pairs	Number of Validation Samples	Classification Accuracy	CPU (min) for Training in Sun Workstation 4/490
Shear	9	24	54.2%	214
Longitudinal	9	24	70.8%	9.3
Shear	9	61	41.0%	150
Longitudinal	9	58	41.4%	6.1
Longitudinal	9	100	57.0%	28.5

* Adapted from Whittaker et al. (1991). With permission.

Table 5.2 Classification of Marbling Levels with Adaptive Logic Networks*

Probe and Input Parameters	Classification Accuracy				Mean Error	CPU (min) for Training in Sun Sparcstation 1
	<3% Fat	3%–7% Fat	>7% Fat	Overall		
Longitudinal with all seven frequency parameters	50.0	61.5	57.1	58.3	1.93	11.9
Shear with all seven frequency parameters	100	53.8	57.1	62.5	1.81	20.0
Longitudinal with the frequency parameters f_c, f_{sk}, and Lm	75.0	61.5	42.0	58.3	2.05	15.7
Longitudinal with the frequency parameters f_c, f_{sk}, and Lm	100	61.5	57.1	58.3	3.85	5.1

* Adapted from Whittaker et al. (1991). With permission.

Table 5.3 Accuracies of Classification of Beef Ultrasonic A-Mode Signals with Unsupervised Training*

Input Parameters	3 Classes of Accuracy	4 Classes of Accuracy	8 Classes of Accuracy
f_a, f_b, f_p, f_c, B^*, f_{sk}, and Lm	68.9%	63.5%	31.1%
f_b, f_p, f_c, B^*, f_{sk}, and Lm	67.6%	54.1%	29.7%
f_p, f_c, B^*, f_{sk}, and Lm	67.6%	54.1%	29.7%
f_c, B^*, f_{sk}, and Lm	66.2%	54.1%	29.7%
B^*, f_{sk}, and Lm	67.9%	58.1%	29.7%
f_{sk}, and Lm	40.5%	23.0%	16.2%

* Adapted from Whittaker et al. (1991). With permission.

A benchmark study was performed to compare these methods using statistical, supervised, and unsupervised ANN training approaches on the basis of independent experiments. In the study, 97 samples with all 7 2.25 MHz shear probe frequency parameters were used for training, and 24 samples extracted from the same data set before training were used for classification. Accuracy was determined in the following ranges: < 3% fat, 3 to 7% fat, and > 7% fat. The number of near misses (NM) was recorded, where a NM was defined as misclassification by ± < 0.5% fat. Table 5.4 shows the results of the benchmark study.

Table 5.4 Benchmark Comparison of Statistical, Supervised,
and Unsupervised ANN Approaches*

| Actual | Predicted Classes | | | | Type II |
Classes	<3%	3 to 7%	>7%	NM	Accuracy
(a) Statistical Regression					
<3%	4	0	0	0	100.0%
3 to 7%	4	8	1	1	61.5%
>7%	0	4	3	3	42.8%
Type I accuracy	50.0%	66.7%	75.0%	—	63.9%
(b) Supervised Training (Adaptive Logic Neural Network)					
<3%	4	0	0	0	100.0%
3 to 7%	6	6	1	2	46.1%
>7%	2	2	3	0	42.9%
Type I accuracy	33.0%	75.0%	75.0%	—	54.2%
(C) Unsupervised Training (Kohonen Self-Organizing Feature Maps Neural Network)					
<3%	3	1	0	0	75.0%
3 to 7%	2	8	3	1	61.5%
>7%	1	4	2	0	28.6%
Type I accuracy	50.0%	61.5%	40.0%	—	55.5%

* Adapted from Whittaker et al. (1991). With permission.

5.1.2 Example: Electronic nose data classification for food odor pattern recognition

By recalling Eq. (4.29), obviously, the classification modeling of food odor by an electronic nose is a multivariate problem with high dimensionality because usually $n \gg 1$, similar to AromaScan's $n = 32$. It usually happens that in a multivariate problem with high dimensionality, the variables are partly correlated. So, a technique is needed to reduce the dimensionality and allow the information to be retained in much fewer dimensions. Principal Component Analysis (PCA) (Hotelling, 1933) is such a linear statistical technique. Through the manipulation of the PCA, a high dimensional data set, such as $\tilde{x}_i$ ($i = 1, 2,..., n$), can be converted into a new data set, such as $\tilde{X}_i$ ($i = 1, 2,..., n$), which are uncorrelated. From the new data set, the first two or three variables are usually good enough to get good model, that is,

$$c = \beta_0^p + \beta_1^p \tilde{X}_1 + \beta_2^p \tilde{X}_2$$

or

$$\hat{c} = \hat{\beta}_0^p + \hat{\beta}_1^p \tilde{X}_1 + \hat{\beta}_2^p \tilde{X}_2 + \hat{\beta}_3^p \tilde{X}_3$$

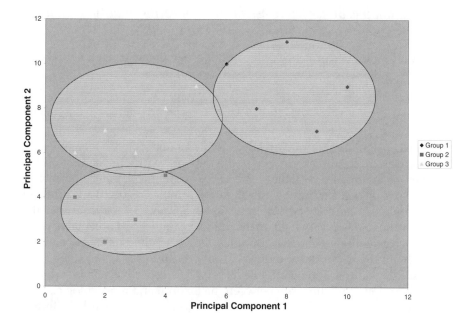

Figure 5.1 Generic plot of an electronic nose data differentiation for three groups of data by the first two principal components.

where $\hat{\beta}_i^p$ ($i = 0, 1, 2$) are the coefficient estimates of the new model. These equations significantly simplify Eq. (4.29). Figure 5.1 shows generically the relationship between three groups of data after principal component processing before the three groups of data may be mixed up together. After the processing, the first two principle components may be used to differentiate the data sufficiently. When the groups of data can be differentiated linearly, that is, they are "linearly separable," the discriminant function can be fitted statistically or with single-layer perceptron neural networks (Rosenblatt, 1959): Otherwise advanced ANN architectures are needed to handle the cases of nonlinear separable or nonlinear. Similarly, if the first three principle components are used to differentiate the data, three dimensional co-ordinates are needed to visualize the data with similar linear separable or nonlinear separable cases handled by statistical and ANN methods.

PCA is the most widely used method for electronic nose data classification. There are, also partial least squares (PLS) (Wold, 1966), cluster analysis (Gardner and Bartlett, 1992), discriminant function analysis (Gardner and Bartlett, 1992), and so on. PLS is especially effective for small sample problems (Yan et al., 1998). Cluster analysis is an unsupervised pattern recognition method, which is self-organized. It is often used together with PCA to identify groups or clusters of points in configuration space (Gardner and Bartlett, 1992). Discriminant function analysis assumes that the data are normally distributed, which limits the use of the method (Gardner and Bartlett, 1992).

5.1.3 Example: Snack food classification for eating quality evaluation based on linear statistical and ANN models

The classification performance of the back propagational trained neural network was judged by defining the classification rate as

$$\text{Classification rate \%} = \frac{NC}{N} \times 100 \tag{5.2}$$

where NC is the number of correctly classified samples.

The classification performance of the network varies depending on the number of features used. Tables 5.5 and 5.6 show the classification rate of the network on some training and validation samples respectively, using all 22 textural, size, and shape features. Tables 5.5 and 5.6 indicate that the classification rate was very high on the training samples and acceptably good on the corresponding validation samples. However, the classification of the quality with the network for reduced 11 and 8 features was not so efficient as the full 22 features, shown in Tables 5.7 and 5.8, respectively. This supports the assumption of nonlinearity between the textural and morphological features and sensory panel scores. Stepwise regression only can help find a

Table 5.5 Performance (% Classification Rate) of Neural Networks with 9 Hidden Nodes on Validation Samples with All 22 Features*

Machine Wear/Raw Material Conditions	Quality/Sensory Attributes						
	Bubble	Rough	Cell	Firm	Crisp	Tooth	Grit
A	92	90	83	88	85	84	94
B	90	98	98	98	94	94	92
C	90	94	86	78	82	82	90
D	90	93	78	95	87	88	90

* Adapted from Sayeed et al. (1995). With permission.

Table 5.6 Performance (% Classification Rate) of Neural Network with 9 Hidden Nodes on Training Samples with All 22 Features*

Machine Wear/Raw Material Conditions	Quality/Sensory Attributes						
	Bubble	Rough	Cell	Firm	Crisp	Tooth	Grit
A	96	98	90	94	93	94	97
B	91	91	98	97	96	89	93
C	95	95	94	88	92	91	97
D	99	100	96	96	99	97	98

* Adapted from Sayeed et al. (1995). With permission.

Table 5.7 Performance (% Classification Rate) of Neural Networks with Reduced 11 Features*

Machine Wear/Raw Material Conditions	Quality/Sensory Attributes						
	Bubble	Rough	Cell	Firm	Crisp	Tooth	Grit
A	81	86	78	65	59	69	88
B	82	94	94	82	82	74	80
C	72	78	60	56	54	48	80
D	82	94	62	75	75	46	75

* Adapted from Sayeed et al. (1995). With permission.

Table 5.8 Performance (% Classification Rate) of Neural Networks with Reduced 8 Features*

Machine Wear/Raw Material Conditions	Quality/Sensory Attributes						
	Bubble	Rough	Cell	Firm	Crisp	Tooth	Grit
A	84	90	76	73	71	71	87
B	96	100	96	96	88	96	92
C	68	84	50	54	68	62	74
D	84	92	73	92	76	80	86

* Adapted from Sayeed et al. (1995). With permission.

compact linear relationship between input and output variables while the reduced features were derived by the stepwise regression which could not identify the nonlinearity. This confirms that ANNs were able to model the nonlinear relationship between the image textural and morphological features and sensory attributes.

The results of this work indicate that the combination of textural and morphological image features can be employed to quantify the sensory attributes of snack quality with a high degree of accuracy from the ANN classifier when compared with human experts.

5.1.4 Example: Meat attribute prediction based on linear statistical and ANN models

Wavelet decomposition as a promising alternative for textural feature extraction from beef elastograms performed much better than the Haralick's statistical method for extraction of textural features in the statistical modeling. This conclusion was based on the prediction ability of the models in terms of the feature parameters extracted by one of the two methods. Huang et al. (1997) showed that the relationship between Haralick's statistical textural feature parameters from beef elastograms and the beef attribute parameters were not significant in the sense of linear statistics. Wavelet textural features from beef elastograms were more informative, consistent, and compact and were used to build models with the ability to predict the attribute parameters acceptably.

Further, Huang et al. (1998) explored the relationship between the wavelet textural features from beef elastograms and the attribute parameters of the beef samples. The purpose of prediction was using the one-hidden-layered feedforward neural networks trained by different methods to implement the process of BP. When compared to the regular BP using the gradient descent method, adding a momentum term improved the training efficiency, and the training epochs were reduced (0.03 to 7.22 times). The Levenberg–Marquardt algorithm was less efficient than the gradient descent algorithm for the cases with reduction of training epochs by 100 times or less. However, it was more efficient for the cases with reduction of training epochs greater than a few hundred times. In the case of difficult convergence in the SARC model using the gradient descent algorithm, the Levenberg–Marquardt algorithm converged much more efficiently. In all cases, the Levenberg–Marquardt algorithm achieved better model output variation accounting and network generalization for attribute prediction. If the consideration of training efficiency were not necessary, the Levenberg–Marquardt algorithm would be a good choice. Further, incorporating the weight decay vs. implementing the Levenberg–Marquardt algorithm alone was effective in improving network generalization resulting in higher R^2 and lower validation MSE values.

This study concluded that ANNs were effective in the prediction of beef quality using wavelet textural feature parameters of the ultrasonic elastograms. ANNs can capture some unknown nonlinear relation between the process inputs and outputs and effectively model the variation in the textural feature space.

5.2 One-step-ahead prediction

In food quality prediction, the purpose of modeling is different for different applications. Controller design for process control is performed based on dynamic process models. In food quality process control, controller design is performed based on prediction from process dynamic models. Some prediction models are for one-step-ahead prediction. Others are for multiple-step-ahead prediction. Therefore, a process model functions as a one-step-ahead predictor, and another functions as a multiple-step-ahead predictor in respective model predictive control (MPC) loops. A one-step-ahead predictor works in an internal model control (IMC) loop, and a multiple-step-ahead predictor works in a predictive control (PDC) loop. The concepts and designs of IMC and PDC loops are discussed in the next chapter. This section focuses on the concept and implementation of one-step-ahead prediction, and the next section covers multiple-step-ahead prediction.

In general, the NARX Eq. (4.5) can be further written as

$$y(t+1) = f(y(t), y(t-1),..., y(t-p+1), u(t), u(t-1),...,$$
$$u(t-q+1), \Theta) + \varepsilon(t+1) \tag{5.3}$$

The one-step-ahead predictor, $\hat{y}(t+1|t)$, can be established based on the preceding equation under the condition that the process output measurements are already known at all previous time instants. For the linear models, the prediction equation can be developed explicitly. In the case of ARX system modeling, Eq. (5.3) takes the following form

$$y(t+1) = A_1 \cdot y(t) + A_2 \cdot y(t-1) + \cdots + A_p \cdot y(t-p+1)$$
$$+ B_1 \cdot u(t) + B_2 \cdot u(t-1) + \cdots + B_q \cdot u(t-q+1)$$
$$+ \varepsilon(t+1) \tag{5.4}$$

where A_i $(i = 1, 2,..., p)$ and B_i $(i = 1, 2,..., q)$ are coefficient matrices. The model itself takes the form

$$\hat{y}(t+1) = \hat{A}_1 \cdot y(t) + \hat{A}_2 \cdot y(t-1) + \cdots + \hat{A}_p \cdot y(t-p+1)$$
$$+ \hat{B}_1 \cdot u(t) + \hat{B}_2 \cdot u(t-1) + \cdots + \hat{B}_q \cdot u(t-q+1) \tag{5.5}$$

where $\hat{A}_i$ $(i = 1, 2,..., p)$ and $\hat{B}_i$ $(i = 1, 2,..., q)$ are the estimates of A_i $(i = 1, 2,..., p)$ and B_i $(i = 1, 2,..., q)$, respectively. Then, the equation of the one-step-ahead predictor can be derived in the following way

$$\hat{y}(t+1|t) = E(y(t+1)|y(t), y(t-1),..., y(1), u(t), u(t-1),..., u(1))$$
$$= E(A_1 \cdot y(t) + A_2 \cdot y(t-1) + \cdots + A_p \cdot y(t-p+1)$$
$$+ B_1 \cdot u(t) + B_2 \cdot u(t-1) + \cdots + B_q \cdot u(t-q+1)$$
$$+ \varepsilon(t+1)|y(t), y(t-1),..., y(1), u(t-1),..., u(1))$$
$$= \hat{A}_1 \cdot y(t) + \hat{A}_2 \cdot y(t-1) + \cdots + \hat{A}_p \cdot y(t-p+1)$$
$$+ \hat{B}_1 \cdot u(t) + \hat{B}_2 \cdot u(t-1) + \cdots + \hat{B}_q \cdot u(t-q+1) \tag{5.6}$$

where $E(\ )$ represents the statistical expectation.

However, for the NARX models, the equation of the predictor can only be developed with an implicit approach. The representation of the NARX one-step-ahead predictor can be obtained as

$$\hat{y}(t+1|t) = E(y(t+1)|y(t),..., y(1), u(t),..., u(1))$$
$$= E(f(y(t),..., y(t-p+1), u(t),..., u(t-q+1), \Theta)$$
$$+ \varepsilon(t+1)|y(t),..., y(1), u(t),..., u(1))$$
$$= \hat{f}(y(t),..., y(t-p+1), u(t),..., u(t-q+1), \hat{\Theta}) \tag{5.7}$$

From this equation, the output of the predictor is the output of the NARX model at time instant $t + 1$, that is, $\hat{y}(t + 1|t) = \hat{y}(t + 1)$. This nonlinear one-step-ahead prediction equation can be adapted by the feedforward neural networks. In this way, a one-step-ahead predictor can be established by a MFN. Eq. (5.7) actually showed that the output of a feedforward network is the one-step-ahead prediction at the time instant t. When, through training, an ANN process model in the structure of MFN is built, the model can be extended to perform one-step-ahead prediction.

5.2.1   Example: One-step-ahead prediction for snack food frying process control

Bullock (1995) presented plots of the predictive ability of the linear ARX model for the snack food frying process. Figure 5.2 shows the ability of the linear ARX model to predict the color and moisture content one-step-ahead on the validation data. It is apparent that the model is highly accurate for

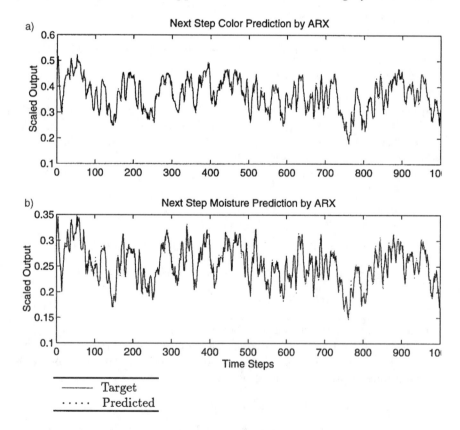

Figure 5.2 One-step-ahead prediction by the linear ARX model for the snack food frying process. (From Bullock, 1995. With permission.)

Table 5.9 MSEs of the 2 × 2 MIMO ANN One-Step-Ahead
Predictions for the Snack Food Frying Process*

Prediction Output	Prediction MSE on Training	Prediction MSE on Validation
Color	0.023177	0.038876
Moisture content	0.006750	0.011718

* From Huang et al. (1998a). With permission.

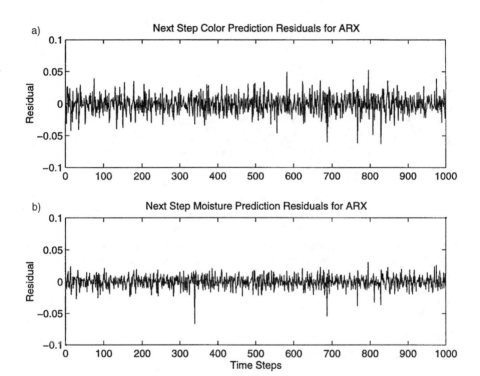

Figure 5.3 Residuals from the one-step-ahead prediction by the linear ARX model for the snack food frying process. (From Bullock, 1995. With permission.)

one-step-ahead prediction because the output of the model is mostly indistinguishable from the actual output. Figure 5.3 shows the plot of the residual of the model.

With the ANN model identification described in the last chapter (Section 4.3.4), Table 5.9, Figure 5.4, and Figure 5.5 show the MSEs and plots of one-step-ahead predictions of the MIMO ANN one-step-ahead predictor. Each of the predictions appears to be exceptionally good so that the ANN prediction model is ready for the design of the IMC loop.

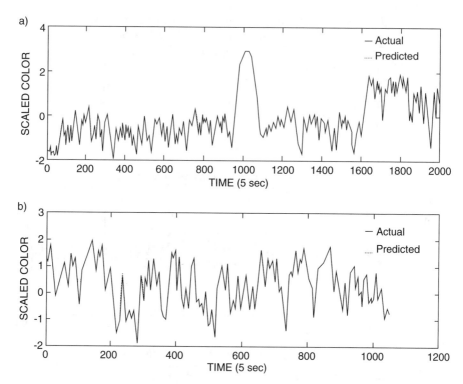

Figure 5.4 (a) One-step-ahead prediction of color from the 2 × 2 MIMO ANN one-step-ahead predictor on training data from the snack food frying process. (b) One-step-ahead prediction of color from the 2 × 2 MIMO ANN one-step-ahead predictor on validation data from the snack food frying process. (From Huang et al., 1998a. With permission.)

5.3 Multiple-step-ahead prediction

Based on the assumption of Eq. (4.5), a feedforward type model is produced which is used in modeling

$$\hat{\underline{y}}(t) = \hat{f}(\underline{y}(t-1), \underline{y}(t-2),\dots, \underline{y}(t-p), \underline{u}(t-1), \underline{u}(t-2),\dots, \underline{u}(t-q), \hat{\Theta})$$

$$(5.8)$$

This model indicates that there is no output feedback to the model input with this modeling approach. That is why this equation can be established by MFNs for one-step-ahead prediction. A different modeling approach is also needed in process control. This approach approximates the general nonlinear function $f(\ )$ in a different way. In this approach, at model input the process outputs at different past instant, $\underline{y}(t-1), \underline{y}(t-2),\dots, \underline{y}(t-p)$,

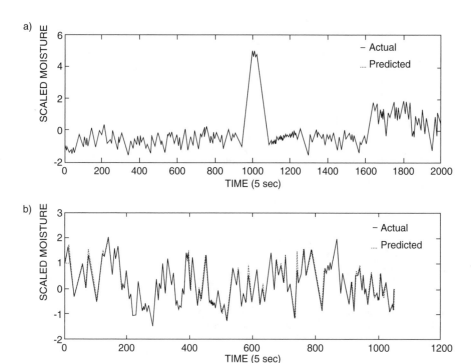

Figure 5.5 (a) One-step-ahead prediction of moisture content from the 2 × 2 MIMO ANN one-step-ahead predictor on training data from the snack food frying process. (b) One-step-ahead prediction of moisture content from the 2 × 2 MIMO ANN one-step-ahead predictor on validation data from the snack food frying process. (From Huang et al., 1998a. With permission.)

are replaced by the model outputs at different past instant, $\hat{y}(t-1)$, $\underline{\hat{y}}(t-2), \ldots, \underline{y}(t-p)$

$$\underline{\hat{y}}(t) = \hat{\underline{f}}\left(\underline{\hat{y}}(t-1), \underline{\hat{y}}(t-2), \ldots, \underline{\hat{y}}(t-p), \underline{u}(t-1), \underline{u}(t-2), \ldots, \underline{u}(t-q), \hat{\Theta}\right)$$

(5.9)

This modeling approach feeds the model output back to the model input. It provides the basis to establish the structure for a multiple-step-ahead predictor, useful in the loop of PDC.

The one-step-ahead predictor can be used to make multiple-step-ahead predictions. The approach is to iterate the one-step-ahead predictor as described in Eq. (5.7), that is, the predictor is to be chained to itself to go as far as needed into the future through successive substitutions of previous predictions. For example, if a predicted output, $\hat{y}(t+l|t)$ $(1 < l < L)$, is needed at a time instant t, then, in the computation, the predicted output, $\underline{\hat{y}}(t+1|t)$, has to replace the actual output, $\underline{y}(t+1)$, measured in the process.

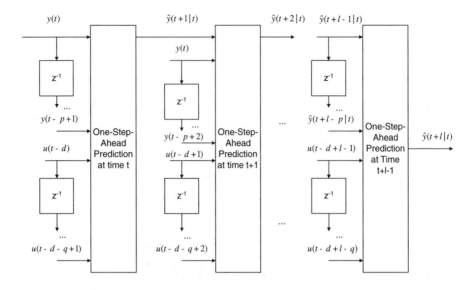

Figure 5.6 A one-step-ahead prediction chain for a multiple-step-ahead prediction. (Adapted from Huang, 1995. With permission.)

Then, $\hat{y}(t+2|t)$ replaces $y(t+2)$ all the way to $t+L-1$, because the actual process outputs in the future are not available yet. Figure 5.6 shows the scheme of such a one-step-ahead prediction chain for multiple-step-ahead prediction. This scheme can be expressed mathematically as

$$\hat{\underline{y}}(t+l|t) = \hat{f}\big(\hat{\underline{y}}(t+l-1|t),\ldots, \hat{\underline{y}}(t+l-p|t), \underline{u}(t+l-1),\ldots,$$

$$\underline{u}(t+l-q), \hat{\Theta}\big) \tag{5.10}$$

$$\hat{\underline{y}}(t+l|t) = \begin{cases} \hat{\underline{y}}(t+l)|l > 0 \\ \underline{y}(t+l)|l \leq 0 \end{cases} \tag{5.11}$$

Realizing the scheme for multiple-step-ahead prediction in linear case is not difficult. The following equation can be derived by chaining Eq. (5.6) to produce the prediction:

$$\hat{\underline{y}}(t+l|t) = \hat{A}_1 \cdot \hat{\underline{y}}(t+l-1|t) + \hat{A}_2 \cdot \hat{\underline{y}}(t+l-2|t) + \cdots + \hat{A}_p \cdot \hat{\underline{y}}(t+l-p|t)$$

$$+ \hat{B}_1 \cdot \underline{u}(t+l-1) + \hat{B}_2 \cdot \underline{u}(t+l-2) + \cdots + \hat{B}_q \cdot \underline{u}(t+l-q) \tag{5.12}$$

However, in the nonlinear case, if you want to use a MFN to realize multiple-step-ahead prediction, a problem occurs. Figure 5.7 shows the multiple-step-ahead (number-of-samples-step-ahead) predictions of quality indices in the snack food frying process described in the example. These predictions were made by the feedforward one-step-ahead predictor.

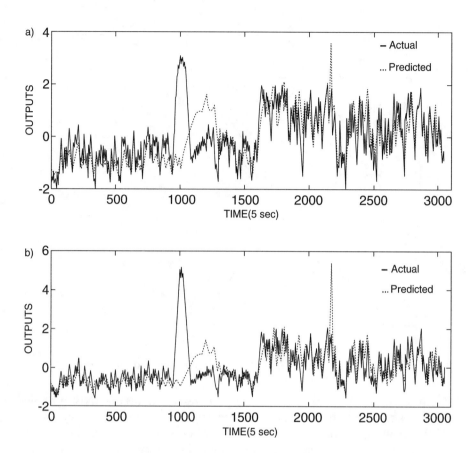

Figure 5.7 (a) Number-of-samples-step-ahead prediction of color with the 2×2 MIMO ANN one-step-ahead predictor. (b) Number-of-samples-step-ahead prediction of moisture content with the 2×2 MIMO ANN one-step-ahead predictor. (From Huang, 1995. With permission.)

These results are not acceptable. Obviously, the one-step-ahead prediction chained multiple-step-ahead predictions are poor. The reason is that the chain of the one-step-ahead predictor resulted in a feedforward network being used as an external recurrent network while the feedforward network was not trained to make multiple-step-ahead prediction. Because of the inherent difference between $y(t)$ and $\hat{y}(t)$, the prediction error can accumulate during iteration and a large error can occur. Strictly speaking, ANN one-step-ahead predictor should not be used to make multiple-step-ahead prediction because the feedforward network is trained to make single-step-ahead predictions only. A multiple-step-ahead prediction needs predictor chaining while the training of the feedforward network does not take such chaining into account. Therefore, the one-step-ahead predictor cannot be used to make reliable multiple-step-ahead prediction in the sense of ANNs. It is necessary to establish an ANN multiple-step-ahead predictor specifically.

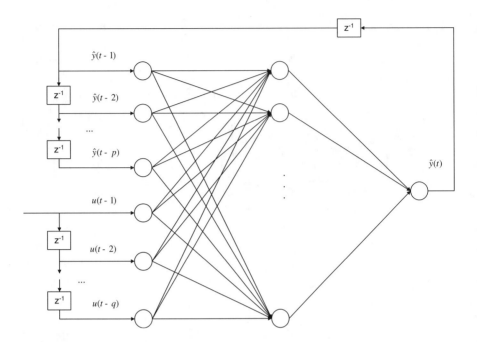

Figure 5.8 Architecture of ERECN with one-hidden layer and one output.

Based on the preceding discussion, the representation of the ANN multiple-step-ahead prediction according to Eqs. (5.10) and (5.11) results in a different ANN architecture compared with the MFN one-step-ahead predictor. This ANN architecture is different from the standard feedforward network because it feedforwards the network input signals to the network output inside (internal) the network and feeds the network output signals back to the network input outside (external) the network. This kind of ANN is an external recurrent neural network (ERECN). Figure 5.8 shows the architecture of the ERECN with one hidden layer and one output. Obviously, the training of this network should be different from the standard feedforward network.

In order to explain the training algorithm of the ERECN for NARX process modeling and prediction, let us discuss some detail of the training algorithm of the MFN first, and then go to the one for the ERECN. For simplicity, the discussion is based on a SISO system assumption. The extension of the concepts and equations to the MIMO system is straightforward.

The training of the feedforward neural networks for one-step-ahead prediction is an optimization problem. The weights in a feedforward network can be adapted to minimize the squared errors as

$$J = \frac{1}{2}\sum_{t=1}^{N}(y(t) - \hat{y}(t))^2 \qquad (5.13)$$

Using the gradient descent algorithm, Rumelhart et al. (1986) developed the BP training algorithm or so-called general delta rule, in which the network weights are updated in proportion to the gradient of J with respect to the weights. In the last chapter, the results of Rumelhart et al. were presented. Earlier Werbos (1974) gave a different derivation of the BP algorithm, which is more general and mathematically more rigorous than the one given by Rumelhart et al. In Werbos' derivation, the chain rule is expressed in a convenient way by ordered derivatives. We are presenting the Werbos' derivation for the BP algorithm for process modeling and prediction. In training, each weight in the network can be updated iteratively

$$w_{ij}^{\text{new}(l)} = w_{ij}^{\text{old}(l)} + \Delta w_{ij}^{\text{old}(l)} \tag{5.14}$$

$$\Delta w_{ij}^{\text{new}(l)} = -\eta \frac{\partial J}{\partial w_{ij}^{\text{old}(l)}} + \rho \Delta w_{ij}^{\text{old}(l)} \tag{5.15}$$

where l represents the number of a layer in the network where $l = 0$ represents the input layer, $l = 1$ represents the hidden layer, and $l = 2$ represents the output layer, and η and ρ are the learning rate and the momentum rate, respectively. The following relationships of ordered (total) derivatives can be derived from the chain rule

$$\frac{\partial^* J}{\partial w_{ij}^{(l)}} = \sum_{t=1}^{N} \frac{\partial^* J}{\partial \text{net}_i^{(l)}(t)} \frac{\partial \text{net}_i^{(l)}(t)}{\partial w_{ij}^{(l)}} \tag{5.16}$$

$$\frac{\partial^* J}{\partial \text{net}_i^{(l)}(t)} = \frac{\partial^* J}{\partial o_i^{(l)}(t)} \frac{\partial o_i^{(l)}(t)}{\partial \text{net}_i^{(l)}(t)} \tag{5.17}$$

$$\frac{\partial^* J}{\partial o_i^{(l)}(t)} = \frac{\partial J}{\partial o_i^{(l)}(t)} + \sum_{k=1}^{n^{(l+1)}} \frac{\partial^* J}{\partial \text{net}_k^{(l+1)}(t)} \frac{\partial \text{net}_k^{(l+1)}(t)}{\partial o_i^{(l)}(t)} \tag{5.18}$$

$$\frac{\partial J}{\partial o_i^{(l)}(t)} = 0 \qquad (l \neq 2) \tag{5.19}$$

where "$*$" represents ordered derivatives, $\text{net}_i^{(l)}(t)$ represents the net output of the ith node in the lth layer at time t, $o_i^{(l)}(t)$ represents the output value after the transfer function at the ith node in the lth layer at time t, and $n^{(l)}$ is the number of nodes in the lth layer. These conventional partial derivatives can be calculated by differentiating the function J with the network equations

$$\frac{\partial \operatorname{net}_i^{(l)}(t)}{\partial w_{ij}^{(l)}} = o_i^{(l-1)}(t) \tag{5.20}$$

$$\frac{\partial o_i^{(l)}(t)}{\partial \operatorname{net}_j^{(l)}(t)} = o_i'^{(l)}(t) \tag{5.21}$$

$$\frac{\partial \operatorname{net}_i^{(l)}(t)}{\partial o_j^{(l-1)}(t)} = w_{ij}^{(l)} \tag{5.22}$$

Define

$$\delta_i^{(l)}(t) = \frac{\partial^* J}{\partial \operatorname{net}_i^{(l)}(t)} \tag{5.23}$$

So

$$\frac{\partial^* J}{\partial w_{ij}^{(l)}} = \sum_{t=1}^{N} \delta_i^{(l)}(t) o_j^{(l-1)}(t) \tag{5.24}$$

This equation can be explained as the change of weights between the $(l-1)$th and the lth layers is determined by the product of the lth layer's δ and the $l-1$st layer's output. Therefore, it is crucial to calculate the δs for updating weights in training. The calculation can be carried out as

$$\delta_i^{(l)}(t) = \frac{\partial^* J}{\partial o_i^{(l)}(t)} o_i'^{(l)}(t) \tag{5.25}$$

$$\frac{\partial^* J}{\partial o_i^{(l)}(t)} = \frac{\partial J}{\partial o_i^{(l)}(t)} + \sum_{k=1}^{n^{(l+1)}} \delta_k^{(l+1)}(t) w_{ki}^{(l+1)} \tag{5.26}$$

The preceding equations contribute to propagating information about error backwards from the $(l+1)$th layer to the lth layer through the δs recursively. For the training of the feedforward network described in the last chapter, these equations become exactly the same as the δ in Eq. (4.35) as long as $\delta^{(2)}(t) = \delta$, $\delta_i^{(1)}(t) = \delta_i$, $o^{(2)}(t) = S_o( )$, and $o^{(1)}(t) = S_h( )$.

For developing the training algorithm of the ERECN for NARX process modeling and prediction, the input to the ERECN can be arranged in the following vector

$$\underline{x}(t) = \hat{y}(t-1), \hat{y}(t-2),\ldots, (\hat{y}(t-p), u(t-1), u(t-2),\ldots, u(t-q))^T$$

The time-delayed output from the network model itself is in this vector $\hat{y}(t-i)$ $(i = 1, 2,\ldots, p)$. In fact, this network model is a specific form of time-lag recurrent networks (Su et al., 1992). Several training algorithms have been proposed for the time-lag recurrent networks with only one recurrent connection, that is, $i = 1$ (Williams and Zipser, 1989; Pineda, 1989; Pearlmutter, 1990). Among the algorithms, the back propagation through time (BPTT) can be modified for multiple recurrent connections, that is, $i > 1$ using ordered derivatives (Werbos, 1974). Actually, when the external feedback signal $\hat{y}(t)$ replaces $y(t)$ as the input to the network, the change of weights will affect $\hat{y}(t+1)$ and, thus, affect $\hat{y}(t+2)$ all the way to $\hat{y}(N)$. The summation term in Eq. (5.26) has to account for this chaining from $t = 0$ to $t = N$. The input layer at time $t+1$ can be considered as a part of the third layer of the network at time t. When calculating the δs of the output layer at time t, it is necessary to calculate the δ for the input nodes at time $t+1$ up to $t+p$, all of which are connected to the corresponding output node at time t. It can be shown that

$$\delta_i^{(0)}(t) = \sum_{j=1}^{p+q} \delta_j(t)w_{ij} \tag{5.27}$$

The δ for the network nodes becomes

$$\delta(t) = [y(t)-\hat{y}(t)] - \sum_{\tau=1}^{p} \delta_{p(\tau-1)+i}^{(0)}(t+\tau)S_o' \left(w_0 + \sum_{j=1}^{h} w_j z_j(t)\right) \tag{5.28}$$

In this equation, the δ propagates the required information all the way from $t = N$ back to the current time instant. This is what BPTT signifies.

This algorithm is used over the whole trajectory of the training data. In training and validation, the number of prediction steps is equal to the total number of training or validation samples. In controller design, the prediction steps are usually specified as a much smaller number than the number of samples. There are two ways to solve this problem. One is to use the long-term predictor trained with the preceding algorithm to do short-term prediction for model predictive control. The other is to train the network to do short-term prediction with a modified training algorithm. Huang (1995) suggested modifying the standard BPTT algorithm in a way that each time instant propagates the δ from the specified prediction step $t + L$ $(L \ll N)$ instead of N back to the current time instant t. It is noted that this modified algorithm requires more memory and computational capability.

5.3.1   Example: Multiple-step-ahead prediction for snack food frying process control

Bullock (1995) demonstrated the ability of the linear ARX model to provide a reliable multiple-step-ahead prediction of the snack food frying process. In the performance test, the models were given starting points and then fed a continuous input of inlet temperature $u_1(t)$ and exposure time $u_2(t)$. The outputs at each time step were fed back in as inputs for the next time step. Figure 5.9 shows the multiple-step-ahead prediction ability of the ARX model. Figure 5.10 shows the residuals of the prediction.

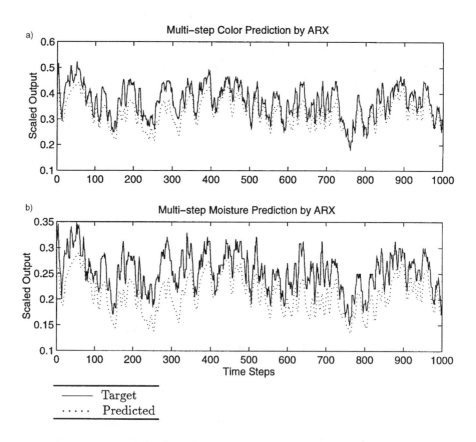

Figure 5.9 Number-of-samples-step-ahead prediction of the ARX model for the snack food frying process. (Adapted from Bullock, 1995. With permission.)

Table 5.10 Predictive MSEs of One- and Multiple-Step-Ahead Predictions
of the Linear ARX Model for the Snack Food Frying Process*

Output	One-Step-Ahead Prediction MSE	Multiple-Step-Ahead Prediction MSE
Color	0.000183	0.005556
Moisture Content	0.000083	0.004578

* Adapted from Bullock (1995). With permission.

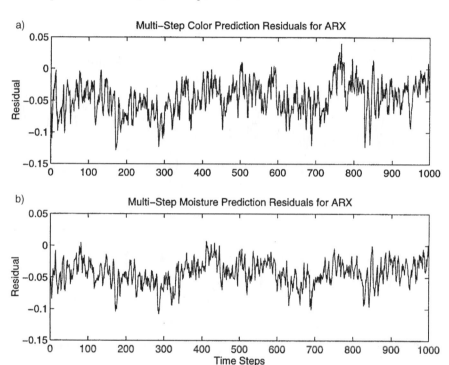

Figure 5.10 Residuals from the number-of-samples-step-ahead prediction of the ARX model for the snack food frying process. (Adapted from Bullock, 1995. With permission.)

It appears that the linear ARX model is undershooting the desired values for the multiple-step-ahead-prediction. In fact, the ARX model was developed to provide one-step-ahead prediction only, while multiple-step-ahead prediction was not integrated into its development.

The results indicate that for the snack food frying process, the linear ARX model did an excellent job for one-step-ahead prediction and a fairly good job for multiple-step-ahead prediction. Table 5.10 shows the error measurement of the ARX model for both one- and multiple-step-ahead predictions for comparison.

Table 5.11 MSEs of the 2 × 2 MIMO ANN Number-of-Samples-Step-
Ahead Predictions for the Snack Food Frying Process*

Prediction Output	Prediction MSE on Training	Prediction MSE on Validation
Color	0.048469	0.060398
Moisture Content	0.048433	0.060044

* From Huang et al. (1998a). With permission.

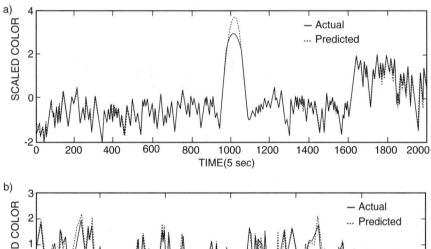

Figure 5.11 (a) Number-of-samples-step-ahead prediction of color from the 2 × 2 MIMO ANN multiple-step-ahead predictor on training data from the snack food frying process. (b) Number-of-samples-step-ahead prediction of color from the 2 × 2 MIMO ANN multiple-step-ahead predictor on validation data from the snack food frying process. (From Huang et al., 1998a. With permission.)

With the training algorithm just described and the ANN model structure parameters (model orders and number of hidden nodes) presented in the Section 4.3.4 for the purpose of comparison, Table 5.11 and Figures 5.11, 5.12 show the MSEs and plots of number-of-sample-steps-ahead predictions of the MIMO ANN multiple-step-ahead predictor. Although the necessary prediction steps would be much smaller, they are all adequate for use as the basis for the design and implementation of the PDC loop.

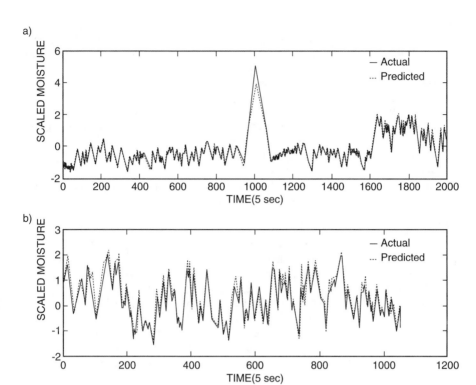

Figure 5.12 (a) Number-of-samples-step-ahead prediction of moisture content from the 2 × 2 MIMO ANN multiple-step-ahead predictor on training data from the snack food frying process. (b) Number-of-samples-step-ahead prediction of moisture content from the 2 × 2 MIMO ANN multiple-step-ahead predictor on validation data from the snack food frying process. (From Huang et al., 1998a. With permission.)

References

Bullock, D., Modeling of a continuous food process with neural networks, M.Sc. thesis, Texas A&M University, College Station, TX, 1995.

Gardner, J. W. and Bartlett, P. N., *Sensors and Sensory Systems for an Electronic Nose.* Kluwer, Dordrecht, The Netherlands, 1992.

Hotelling, H., Analysis of a complex of statistical variables into principal components, *J. Educ. Psychol.,* 24, 498, 1933.

Huang, Y., Snack food frying process input–output modeling and control through artificial neural networks, Ph.D. dissertation. Texas A&M University: College Station, TX, 1995.

Huang, Y., Lacey, R. E., Moore, L. L., Miller, R. K., Whittaker, A. D., and Ophir, J., Wavelet textural features from ultrasonic elastograms for meat quality prediction, *Trans. ASAE,* 40(6), 1741, 1997.

Huang, Y., Lacey, R. E., and Whittaker, A. D., Neural network prediction modeling based on elastographic textural features for meat quality evaluation, *Trans. ASAE,* 41(4), 1173, 1998.

Huang, Y., Whittaker, A. D., and Lacey, R. E., Neural network prediction modeling for a continuous, snack food frying process, *Trans. ASAE,* 41(5), 1511, 1998a.

Pearlmutter, A. B., Dynamic recurrent neural networks. in Tech. Rep. CMU-CS-90-196, Carnegie Mellon University, Pittsburgh, PA, 1990.

Pineda, F. J., Recurrent backpropagation and the dynamical approach to adaptive neural computation, *Neur. Comp.,* 1, 167, 1989.

Rosenblatt, R., *Principles of Neurodynamics,* Spartan Books, New York, 1959.

Rumelhart, D. E., Hinton, G. E., and Williams, R. J., Learning internal representation by error propagation, in *Parallel Distributed Processing: Explorations in the Microstructures of Cognition,* Vol. I, MIT Press, Cambridge, MA, 1986, chap. 8, 318.

Sayeed, 1995.

Su, H., McAvoy, T. J., and Werbos, P. J., Long-term predictions of chemical processes using recurrent neural networks: a parallel training approach, *Ind. Eng. Chem. Res.,* 31(5), 1338, 1992.

Werbos, P. J., Beyond regression: new tools for prediction and analysis in the behavior sciences, Ph.D. dissertation, Harvard University, Boston, MA, 1974.

Whittaker, A. D., Park, B., McCauley, J. D., and Huang, Y., Ultrasonic signal classification for beef quality grading through neural networks, *Proc. 1991 Symp. Autom. Agric. 21st Century,* ASAE, St. Joseph, MI, 1991.

Williams, R. J. and Zipser, D. A., A learning algorithm for continually running fully recurrent neural networks, *Neur. Comput.,* 1, 270, 1989.

Wold, H., Nonlinear estimation by iterative least squares procedures, in *Research Papers in Statistics,* David, F., Ed., John Wiley & Sons, New York, 1966.

Yan, D., Huang, Y., and Lacey, R. E., Assessment of principal component neural networks for extraction of elastographic textural features in meat attribute prediction, ASAE paper No. 98-3018, St. Joseph, MI, 1998.

chapter six

Control

6.1 Process control

In food process quality control, human operators are subject to overcorrection and overreaction to normal process variability. In order to ensure consistency in product quality on-line, automatic control is desired. For control purposes, it is usually necessary to identify the process model first. Then, based on the model, determine the scheme or controller needed to regulate the process dynamics. The process of model building involves sampling, data acquisition, data dynamic analysis, process modeling, and process prediction. These topics were covered in previous chapters. This chapter focuses on the issues of controller design, simulation, and implementation in food processes.

When an accurate model of a process under consideration is obtained, it can be used in a variety of control strategies. Model predictive control (MPC) is a well-established model-based control structure in control theory and practice. In the MPC family, internal model control (IMC) and predictive control (PDC) have been widely used in practical process control. IMC emphasizes the role of process forward and inverse models. In this structure, the process forward and inverse models are used directly as elements in a designed feedback closed loop, and the difference between the process and model outputs is fed back for regulating the performance of the control system. The IMC approach is strongly supported by control theory. However, it should be noted that the implementation of the IMC structure is limited to open-loop stable systems. Even so, this control structure still has been widely used in process control. The PDC is based on the receding horizon technique. It has been introduced as a natural, computationally feasible, feedback law in the realm of optimal control. In this approach, a process model can be used to provide prediction of the future process response over the specified horizon. The predictions supplied by the process model are passed to a numerical optimization routine that minimizes a designated system performance criterion in the calculation of suitable control signals. An advantage of the PDC approach is that it can include the constraints for process inputs and outputs and model uncertainties in the formulation while IMC does not consider it explicitly.

In the area of model-based process control, perhaps the most significant MPC strategies are IMC and PDC. In this chapter, the basic principles of the two control strategies will be introduced, the algorithms for controller design will be formulated, and the examples from practical food processes will be presented.

6.2   Internal model control

The IMC strategy was first proposed by Garcia and Morari in 1982 in terms of linear SISO systems. Later Economou et al. (1986) extended it to general nonlinear systems. The linear IMC structure is shown in Figure 6.1. In this diagram, the following relationship can be established

$$y = P \times \hat{u} + \varepsilon \tag{6.1}$$

$$\hat{u} = [1 + C(P - M)]^{-1} C(y^s - \varepsilon) \tag{6.2}$$

where P represents the process, C represents the IMC controller, M represents the process model, and ε represents measurement noise from the process. There are three properties of the linear IMC structure (Garcia and Morari, 1982).

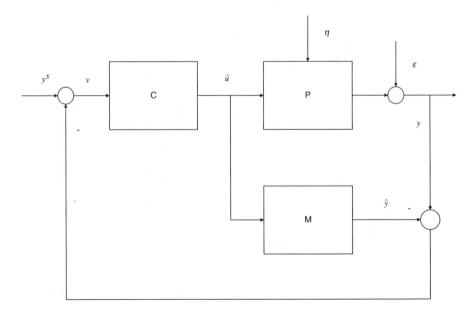

Figure 6.1 Structure of a linear IMC loop. (Adapted from Huang, 1995. With permission.)

1. Dual stability—assume the model is perfect, that is, $M = P$. The closed-loop system in Figure 6.1 is stable if the controller, C, and the process, P, are stable. This implies that unless there are modeling errors and as long as the open-loop system is stable, the stability issue is trivial.
2. Perfect control—assume that the controller is equal to the model inverse, that is, $C = M^{-1}$, and that the closed-loop system in Figure 6.1 is stable. Then, $y(t) = y^s(t)$ for all $t > 0$ and all noises $\varepsilon(t)$. This reasserts that the ideal open-loop controller leads to perfect closed-loop performance in the IMC structure.
3. Zero offset—assume that the steady-state gain of the controller is equal to the inverse of the model gain, that is, $C_\infty = M_\infty^{-1}$, and that the closed-loop system in Figure 6.1 is stable. Then, there will be no offset for asymptotically constant set points and noises, $\lim_{t \to \infty} y(t) = y^s$. Integral-type control action can be built into the structure without the need for additional tuning parameters.

In an analogy to the linear case, the design approach of the IMC controller can be extended to nonlinear systems in an orderly fashion as follows (Economou et al., 1986). The nonlinear IMC structure is shown in Figure 6.2. The blocks with double lines are used to emphasize that the operators are nonlinear and that the usual block diagram manipulations do not hold.

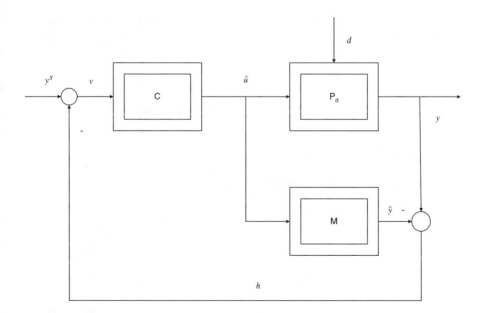

Figure 6.2 Structure of a nonlinear IMC loop. (Adapted from Huang, 1995. With permission.)

The following relationships are from the diagram

$$\hat{y} = M(C(v)) \tag{6.3}$$

$$v = y^s - y + \hat{y} \tag{6.4}$$

$$h = P_d(C(v)) - M(C(v)) + \varepsilon \tag{6.5}$$

The three properties of the nonlinear IMC structure are similar to the three properties of the linear IMC structure (Economou et al., 1986).

1. Stability—if C and P_d are input–output stable and a perfect model of the process is available, that is, $M = P_d$, then the closed-loop system is input–output stable.
2. Perfect control—if the right inverse of the model operator M^i exists, that is, $C = M^i$, and the closed-loop system is input–output stable with this controller, then the control will be perfect, that is, $y = y^s$.
3. Zero offset—if the right inverse of the steady-state model operator M_∞^i exists, that is, $C_\infty = M_\infty^i$, and the closed-loop system is input–output stable with this controller, then offset free control is attained for asymptotically constant inputs.

A very important difference between linear and nonlinear processes exists. For a linear process, noises can be assumed to perform additive actions on the output without loss of generality because of the superposition principle

$$\begin{aligned} P(u + \eta) + \varepsilon &= P \times u + P \times \eta + \varepsilon \\ &= P \times u + \delta \quad \text{with } \delta = P \times \eta + \varepsilon \end{aligned}$$

This principle does not hold for a nonlinear process. In the preceding equation, the symbol P_d is used for the process operator to signify the effect of unmeasurable noises resulting in differences between the model and the process.

The preceding linear and nonlinear IMC structures provide the direct methods for the design of linear and nonlinear feedback controllers.

The procedure for IMC design consists of the following two steps:

1. Design the controller $\hat{C}$ under the assumption of perfect control, that is, $P = M$.
2. A filter F needs to be designed to preserve the closed-loop characteristic that $\hat{C}$ was designed to produce in reality, the control cannot be perfect, and there always exists a mismatch or error between the model and the plant (process).

This procedure can be implemented fairly easily in linear IMC design. For Step 1, $\hat{C} = 1/M$, the controller $\hat{C}$ is the direct inverse of the model M. Then, for Step 2, the final IMC controller C is set by augmenting $\hat{C}$ with F, a low pass filter, so that $C = \hat{C}F$. F is a low pass filter whose parameters, with the ones in $\hat{C}$, can be adjusted to improve the robustness. For example, a common form of the filter used for discrete SISO is first order (Prett and Garcia, 1988)

$$F(z) = \frac{(1 - \delta)}{z - \delta} z$$

where $0 < \delta < 1$ is the filter time constant. This procedure can be implemented in a similar but much more complicated manner in nonlinear IMC design.

ANN is a promising alternative to handling nonlinearity in modeling. It is also a promising alternative for design and implementation of controllers for nonlinear process control. As a general nonlinear control scheme, a neural network needs to be trained first to represent the process response. This network is used as the process model, M, in the IMC structure. Then, a second network is trained to represent the inverse of the process dynamics. This training work can be performed with the structure shown in Figure 6.3. Having obtained the inverse network model, this network is used as the controller, C, in the IMC structure. The inverse of a nonlinear process plays a central role in the development of the nonlinear IMC controllers.

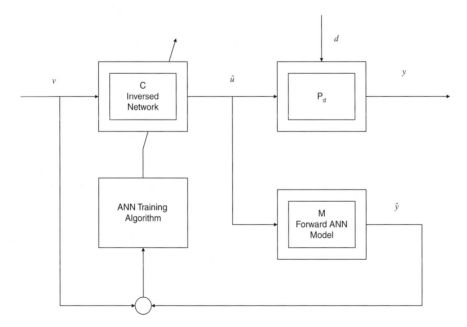

Figure 6.3 Process ANN model inverse training structure. (Adapted from Huang, 1995. With permission.)

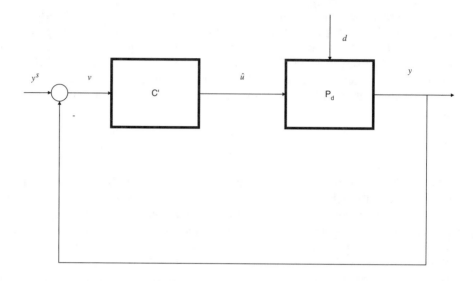

Figure 6.4 Feedback control structure. (Adapted from Huang, 1995. With permission.)

An alternative method for achieving a stable inverse of a neural network process model is to calculate the inverse model iteratively. With this method, the neural network process model equations are treated as nonlinear functions, and a numerical method can be used to compute the inverse of these functions at each sampling instant. This method is effective even for those processes that are only invertible locally, but training an inverse neural network requires the process to be invertible globally, that is, over the whole operating space. In some cases, this may cause difficulty in learning the accurate inverse of the process dynamics through inverse network training. Therefore, the iterative inversing method is more reliable than the method of inverse network training. In this chapter, the iterative inversing method is recommended for the design of an ANN-based IMC (ANNIMC) controller.

It is important to note that the IMC structure is equivalent to the classical feedback structure, as shown in Figure 6.4, in which the bold lines emphasize the general relationship whether they are linear or nonlinear. This equivalence is in the sense that any external inputs, y^s, η, and ε, to the closed-loop system will give rise to the same internal signals, u and y. In the linear case, the IMC controller, C, and the feedback controller, C_f, are related by

$$C = C_f (1 + MC_f)^{-1} \tag{6.6}$$

and

$$C_f = C (1 - MC)^{-1} \tag{6.7}$$

By this equivalence, the properties of the IMC can be explained easily through the concept of feedback control. Whatever is possible with the feedback control is possible with the IMC and vice versa.

In the nonlinear case, the feedback controller, C_f, and the IMC controller, C, are related by

$$\hat{u} = C(v)$$
$$= C_f(e) \tag{6.8}$$

and

$$v = e + \hat{y} \tag{6.9}$$

where $e = y^s - y$. A filter, F, can be employed in series with the IMC controller, C, for the robustness of the controller. In the frequency domain, the preceding relationship of v and e can be further expressed as

$$V(z) = F(z)(E(z) + \hat{Y}(z)) \tag{6.10}$$

In an IMC loop, if the process model is perfect and u can be solved from the model in the form of a one-step-ahead predictor, the following relation can be obtained

$$v(t) = \hat{y}(t+1) \tag{6.11}$$

Taking a z-transform on this equation and combining with Eq. (6.10), the relationship between the error signals of feedback control, e, and of IMC, v, is obtained as

$$V(z) = \frac{F(z)}{1 - F(z)z^{-1}} E(z) \tag{6.12}$$

In this way, a feedback controller and an IMC controller are equivalent on the basis of the process prediction model. When designing and implementing a feedback controller based on the process model, an IMC controller is designed and implemented equivalently.

As mentioned previously, the inverse of the neural network model is crucial in the design of the nonlinear IMC controller. This inverse can be calculated iteratively at each numerical sampling instant. Newton's method and the gradient descent method (Scales, 1985) can be used to develop algorithms for the calculation. Based on the inverse strategy, the ANNIMC loop can be diagramed as shown in Figure 6.5 where $M^{p(1)}$ represents the neural network one-step-ahead prediction model.

As described in the last two chapters, the neural network one-step-ahead prediction model can be written in the form

$$\hat{y}(t+1) = \hat{f}(y(t), y(t-1), \ldots, y(t-p), u(t), u(t-1), \ldots, u(t-q), W) \tag{6.13}$$

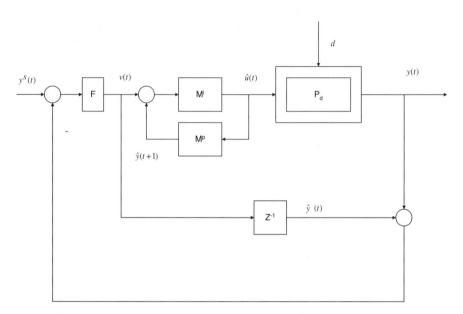

Figure 6.5 Structure of ANNIMC. (Adapted from Huang, 1995. With permission.)

The right inverse of M can be generated by solving the preceding non-linear equation for the current process input $u(t)$. Because $u(t)$ is unknown at this instant, $\hat{y}(t+1)$ cannot be estimated. In this way, the IMC controller, C, is designed by replacing $\hat{y}(t+1)$ with an internal signal $v(t)$ in the control loop and solving the resulting equation for $u(t)$

$$v(t) = \hat{f}(y(t), y(t-1),\ldots, y(t-p), u(t), u(t-1),\ldots, u(t-q), W) \quad (6.14)$$

A function can be further defined as

$$U(u(t)) = v(t) - \hat{f}(y(t), y(t-1),\ldots, y(t-p), u(t), u(t-1),\ldots, u(t-q), W)$$

$$(6.15)$$

The general equation for the iteration of $u(t)$ is

$$\hat{u}^k(t) = \hat{u}^{k-1}(t) + \Delta^{k-1}\hat{u}(t) \quad (6.16)$$

where k represents the number of iteration. In terms of the concept of Newton's method, $\Delta^{k-1}\hat{u}(t)$ in the Eq. (6.19) should be

$$\Delta^{k-1}\hat{u}(t) = -\frac{U(u(t))}{\partial U(u(t))/\partial u(t)}\bigg|_{u(t)=\hat{u}^{k-1}(t)} \quad (6.17)$$

The forms of $U(u(t))$ and $\partial U(u(t))/\partial u(t)$ depend on the chosen neural network structure. They can be derived according to the equations of the specified neural network model.

Eqs. (6.16) and (6.17) with their expansion from the specified neural network model formulate the algorithm of the IMC controller based on a neural network one-step-ahead prediction model with Newton's method. By use of this algorithm, the iterative sequence is initialized by the process input calculated at the previous sampling instant, that is,

$$\hat{u}^0(t) = u(t-1) \tag{6.18}$$

If the algorithm converges after m iterations, the current process input is assigned as

$$u(t) = \hat{u}^m(t) \tag{6.19}$$

In practice, Newton's method is sensitive to initial conditions. If the initial conditions are not perfect, the iterative procedure may not converge. In order to alleviate this problem, a factor simply can be introduced to the iterative Eq. (6.16) to form a modified iterative equation of Newton's method

$$\hat{u}^k(t) = \hat{u}^{k-1}(t) + \gamma \Delta^{k-1} \hat{u}(t) \tag{6.20}$$

where γ is the factor. The factor can control the convergence speed of the algorithm. When $\gamma = 1$, the algorithm is the standard Newton's method, and the convergence speed of the algorithm is fastest when the convergence space is small. When $\gamma \leq 1$, the convergence speed may be slower and the convergence space may be larger which, actually, relaxes the requirement for the initial conditions. On the other hand, in the implementation of the control system because the factor works in the control loop, it has an effect on the controller performance. Therefore, the factor can be used as a controller tuning parameter.

Newton's method is to approximate a nonlinear function with piecewise linearization in an iterative procedure. An alternative method is setting up an objective function

$$J(u(t)) = \frac{1}{2} \sum_{t=1}^{N} U^2(u(t))$$

$$= \frac{1}{2} \sum_{t=1}^{N} (v(t) - \hat{y}(t+1))^2 \tag{6.21}$$

In this way, the problem of solving a nonlinear function is changed to minimizing an objective function. A number of methods can be used to accomplish this. Among these methods, the gradient descent method has been well established. With the method of gradient descent, the iterative equation of $u(t)$ is still similar to Eq. (6.16), but the term $\Delta^{k-1} \hat{u}(t)$ is changed to

$$\Delta \hat{u}(t) = -\gamma \frac{\partial J(u(t))}{\partial u(t)} \bigg|_{u(t)=\hat{u}(t)} \tag{6.22}$$

This expression guarantees that each update of $u(t)$ makes the objective function, J, move in the direction of the greatest gradient. Further,

$$\frac{\partial J(u(t))}{\partial u(t)} = -(v(t) - \hat{y}(t+1))\frac{\partial \hat{y}(t+1)}{\partial u(t)} \qquad (6.23)$$

The form of $\partial \hat{y}(t+1)/\partial u(t)$ also depends on the chosen neural network structure. It can be derived according to the equations of the specified neural network model.

Eqs. (6.16), (6.22), and (6.23) with their expansion from the specified neural network model formulate the algorithm of the IMC controller based on a neural network one-step-ahead prediction model with the gradient descent method. Just as with the implementation of Newton's method-based algorithm, the iterative sequence in the gradient descent method is initialized with the input calculated at the previous sampling instant as in Eq. (6.18). If the algorithm converges after m iterations, the current input is assigned as in Eq. (6.19). The parameter γ has a similar effect as in the equation of the modified Newton's method. It can control the convergence of the algorithm and has an effect on the control loop as well.

When using the modified Newton's method or the gradient descent method to calculate the inverse of the neural network process one-step-ahead prediction model, the controller will perform differently with respect to sluggish tracking, smooth and fast tracking, or oscillation with the change of γ. Therefore, controller tuning is important in controller design. The purpose of tuning is to find the value of a certain parameter in the control loop resulting in satisfactory performance of the controller. The parameter, γ, discussed previously can be tuned to achieve smooth control of the ANNIMC controller with a change of the response. There are two aspects that need to be considered in controller tuning. In considering the strategy of the solution, tuning with γ is actually a one-dimensional search problem. However, because it is difficult, in practice, to guarantee that the criterion is convex throughout the search interval, the direct one-dimensional search method may not be useful. For specific tuning with γ in the ANNIMC loop, the following procedure can be followed.

1. Determine the tuning interval—the two extremes of γ that correspond to sluggish tracking and oscillation can be determined empirically. These two numbers define the interval.
2. Determine the increment of γ—the increment of γ, $\Delta\gamma$, is determined in terms of the sensitivity of the controller performance to γ.
3. Perform the experiment—over the tuning interval, beginning from the left extreme, the controller experiment points are set up according to

$$\gamma^{new} = \gamma^{old} + \Delta\gamma^{old} \qquad (6.24)$$

until the right extreme is reached. The controller is implemented at each experimental point and the selection criterion is calculated at each value.

4. Determine the final γ—choose the final γ, the point where the corresponding controller has the minimum value of the criterion. If there are several possible final γ candidates from different criteria, it may be necessary to compare the corresponding controller performance to make a final decision.

The second aspect is related to the criterion used in tuning a controller. This criterion usually is an objective function that can represent the characteristics of the control system and is convenient in computation. Obviously, when different objective functions are selected for use for the same control system, the final parameters may be different.

In general, there are two types of objective functions. The first is established by directly using the output response characteristics of the control system. Typically, under a step input, the desired output of the control system changes as shown in Figure 6.6.

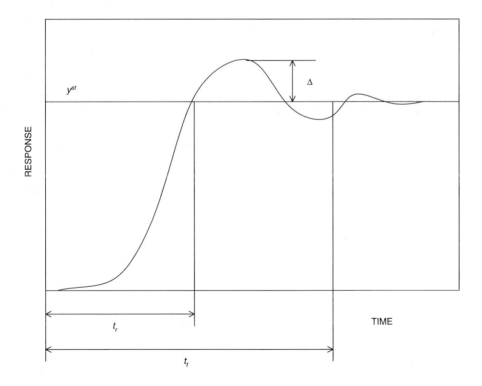

Figure 6.6 Typical output response of a control system under a step input. (From Huang, 1995. With permission.)

The following characteristics are used to establish objective functions.

1. Rising time t_r—this is the time at which the output, $y(t)$, reaches the static value for the first time.
2. Regulating time t_f—this is the time at which the output, $y(t)$, reaches the static area of 95 to 105 percent and never exceeds those limits.
3. Overshoot δ percent

$$\delta\% = \frac{\Delta}{y^{st}} \times 100\% \qquad (6.25)$$

where Δ is the maximum difference in which $y(t)$ deviated from y^{st}, and y^{st} is the static value of $y(t)$.

However, only one of these characteristics is often selected to establish the objective function in tuning, while the other two are used for testing conditions. This method cannot achieve optimal tuning based on all characteristics.

The second type of objective function is called the error objective function. It is based on the difference between the desired system response (usually step response) and the real system response. This is done to manipulate mathematically the three characteristics in the first type of objective function and to integrate them into a single mathematical equation. This type of objective function achieves comprehensive tuning while integrating all characteristics.

Three error objective functions are popular in controller evaluation. They are based on different forms of minimizing

1. The error integral—the minimum integral of square error (ISE),

$$J_{ISE} = \int_0^\infty e^2(t)\, dt \qquad (6.26)$$

2. The minimum integral of absolute error (IAE)

$$J_{IAE} = \int_0^\infty |e(t)|\, dt \qquad (6.27)$$

3. The minimum integral of absolute error multiplied by time (ITAE)

$$J_{ITAE} = \int_0^\infty t|e(t)|\, dt \qquad (6.28)$$

In practice, the integral upper limits in the preceding three equations are set up to be multiples of the transient time of the control system instead of ∞.

6.2.1 Example: ANNIMC for the snack food frying process

The preceding theoretical description needs to be extended further for the development of the IMC controller for the snack food frying process. As shown previously for process quality control, the continuous, snack food frying process can be generalized as a dynamic system with two inputs, the inlet temperature of the frying oil and the residence time of the product staying in the fryer, and two outputs, the product quality attributes of color and moisture content. If the system equation of discrete-time time-delayed, nonlinear, autoregressive with exogenous input (TDNARX) is assumed to govern the 2×2 process (Huang et al., 1998), then this process can be modeled by a feedforward neural network

$$\hat{y}(t) = \hat{f}[y(t-1), y(t-2),..., y(t-p), u(t-d-1),$$
$$u(t-d-2),..., u(t-d-q), W) \tag{6.29}$$

where $y(t) = [y_1(t), y_2(t)]^T$ is the process output vector in which $y_1(t)$ is the product color and $y_2(t)$ is the product moisture content (percent) at time t, $\hat{y}(t) = [\hat{y}_1(t), \hat{y}_2(t)]^T$ is the approximated process output vector by the model, $u(t) = [u_1(t), u_2(t)]^T$ is the process input vector in which $u_1(t)$ is the inlet temperature (°C) and $u_2(t)$ is the residence time (s) at time t, p represents the orders of the past outputs $y_1(t)$ and $y_2(t)$ in the vector equation, that is, $y(t-p) = [y_1(t-p_1), y_2(t-p_2)]^T$ in which p_1 and p_2 are the maximum orders of the past outputs related to the present outputs $y_1(t)$ and $y_2(t)$, respectively. q represents the orders of the past inputs $u_1(t)$ and $u_2(t)$ in the vector equation, that is, $u(t-q) = [u_1(t-q_1), u_2(t-q_2)]^T$ in which q_1 and q_2 are the maximum orders of the past inputs related to the present outputs $y_1(t)$ and $y_2(t)$, respectively, and d represents the time lags from the process input $u(t)$ to the process output $y(t)$, that is, $u(t-d) = [u_1(t-d_1), u_2(t-d_2)]^T$ in which d_1 and d_2 are the minimum time lags of the inputs related to the present outputs $y_1(t)$ and $y_2(t)$, respectively.

This equation can be further rewritten in the form of a one-step-ahead prediction. Equivalently,

$$\hat{y}(t+d+1) = \hat{f}[y(t+d), y(t+d-1),..., y(t+d-p+1),$$
$$u(t), u(t-1),..., u(t-q+1), W] \tag{6.30}$$

This equation contains the input vector, $u(t)$, to be used to compute the control action for a unique solution of $u(t)$. A function can be defined for this purpose

$$U(u(t)) = v(t) - \hat{f}(\hat{y}(t+d), \hat{y}(t+d-1),..., \hat{y}(t+d-p+1),$$
$$u(t), u(t-1),..., u(t-q+1), W) \tag{6.31}$$

where $v(t)$ is the tracking signal of the network model predicted output $\hat{y}(t+d+1)$.

Therefore, the control actions can be computed with the inverse of the function $\underline{U}(\underline{u}(t))$. However, the future process outputs $\underline{y}(t + d - p + 1),...,$ $\underline{y}(t + d)$ need to be estimated in order to realize the computation.

Because the neural network prediction model equation is nonlinear, an analytical solution cannot be obtained. The control law needs to be evaluated by solving the nonlinear function $\underline{U}(\ )$, the process prediction model, at each time instant iteratively in a numerical fashion

$$\hat{\underline{u}}^k(t) = \hat{\underline{u}}^{k-1}(t) + \Delta^{k-1}\hat{\underline{u}}(t) \tag{6.32}$$

where $\hat{\underline{u}}^k(t)$ is the computed $\underline{u}(t)$ at the kth iteration, and $\Delta^{k-1}\hat{\underline{u}}(t)$ is the updating increment of $\hat{\underline{u}}(t)$ at the kth iteration, determined by a certain numerical method.

The process model one-step-ahead prediction and the control action computation set up an ANN process one-step-ahead prediction model-based IMC (ANNIMC) loop for the continuous snack food frying process. The structure of the ANNIMC loop is shown in Figure 6.7 where $\underline{\varepsilon}(t) = (\varepsilon_1(t), (\varepsilon_2(t))^T$ is the measurement noise vector in which $\varepsilon_1(t)$ is for $y_1(t)$ and $\varepsilon_2(t)$ is for $y_2(t)$ at time t, assumed to be two-dimensional Gaussian distributed with zero mean and certain variance, $\underline{\eta}(t)$ represents the system noises, M^p is the neural process one-step-ahead prediction model, M^i is the inverse of the neural process prediction model, F is a filter for the robustness of the controller, and $\underline{y}^s(t)$ contains the setpoints of the process outputs.

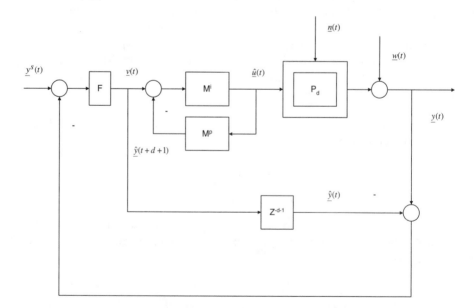

Figure 6.7 Structure of ANNIMC for the snack food frying process. (Adapted from Huang et al., 1998a. With permission.)

As previously discussed, the inverse of the neural network process one-step-ahead prediction model is crucial in setting up an IMC loop. This inverse computes the control actions iteratively at each time instant. For snack food frying process control, Newton's method and the gradient descent method were applied and compared to obtain the inverse.

According to Newton's method, the term $\Delta^{k-1}\hat{\underline{u}}(t)$ in Eq. (6.32) can be written as

$$\Delta^{k-1}\hat{\underline{u}}(t) = -\underline{U}(\hat{\underline{u}}^{k-1}(t))\left(\frac{\partial \underline{U}(\underline{u}(t))}{\partial \underline{u}(t)}\right)^{-1}\Bigg|_{\underline{u}(t)=\hat{\underline{u}}^{k-1}(t)} \tag{6.33}$$

where $\partial \underline{U}(\underline{u}(t))/\partial \underline{u}(t)$ is a 2×2 Jacobian matrix, and each component can be derived from the equations of the process prediction model (Huang, 1995).

Newton's method is sensitive to initial conditions during iterative computation. If the initial conditions are poor, the iterative procedure may not converge. To alleviate the problem, a factor, γ, can be incorporated into the standard iterative equation of Newton's method to form a modified iterative equation for Newton's method.

$$\hat{\underline{u}}^k(t) = \hat{\underline{u}}^{k-1}(t) + \gamma^T \Delta^{k-1}\hat{\underline{u}}(t) \tag{6.34}$$

The factor, γ, controls the convergence speed of the algorithm. The algorithm is the standard Newton's method when $\gamma = (1, 1)^T$; otherwise, it is the modified Newton's algorithm with larger convergence space and relaxed requirements on the initial conditions. In a control loop, the factor, γ, can be tuned for the performance of the controller.

An alternative way to perform the inverse computation is to use the function $\underline{U}(\underline{u}(t))$ to set up an optimization problem to minimize the following objective function

$$J(\underline{u}(t)) = \frac{1}{2}\underline{U}^T(\underline{u}(t)) \cdot \underline{U}(\underline{u}(t)) \tag{6.35}$$

The control actions can be updated using the gradient descent method

$$\Delta^{k-1}\hat{\underline{u}}(t) = -\gamma^T\frac{\partial J(\underline{u}(t))}{\partial \underline{u}(t)}\Bigg|_{\underline{u}(t)=\hat{\underline{u}}^{k-1}(t)} \tag{6.36}$$

where $\partial \underline{U}(\underline{u}(t))/\partial \underline{u}(t)$ is also a 2×2 Jacobian matrix, in which each component can also be derived from the equations of the neural network process prediction model (Huang, 1995). Similarly, the factor, γ, can also be tuned in the control loop. This expression guarantees that each update of $\underline{u}(t)$ makes the objective function J move in the direction of the largest gradient decrease.

The two time lags from the process inputs, inlet temperature and residence time, are separated 20 and 16 units (each unit accounts for 5 s), respectively, from the process outputs, color, and moisture content. It can be derived that at time, t, the model inverse can only be solved for $\tilde{u}(t) = (u_1(t),$ $u_2(t + 4))^T$ instead of $u(t) = (u_1(t), u_2(t))^T$ (Huang, 1995). In the simulation computation, empirical estimations of future control actions $u_2(t + 3)$, $u_2(t + 2)$, $u_2(t + 1)$, and $u_2(t)$ are needed at the initial stage and, then, the previously computed u_2s are saved to implement $u_1(t)$ and $u_2(t)$ at the same time instant.

During the controller tuning process, $\gamma_1 = \gamma_2 = \tilde{\gamma}$ was assumed. The tuning could be performed in the interval with a parameter increment in a one-dimensional space. The three objective functions, ISE, IAE, and ITAE, test and verify each other (Huang, 1995).

Using the modified Newton's method to compute the inverse of the neural process one-step-ahead prediction model, it was found that as the values of the factor $\tilde{\gamma}$ were increasing, the performance of the controller went from sluggish setpoint tracking, to smooth and fast setpoint tracking, to oscillation.

Table 6.1 shows the tuning parameters where the final $\tilde{\gamma}$ were 0.05 with ISE and 0.04 with IAE or ITAE. The two different responses of color and moisture are plotted in Figure 6.8. The responses with $\tilde{\gamma} = 0.04$ are smoother. Consequently, $\tilde{\gamma} = 0.04$ with IAE and ITAE was preferred. Note that in the controller test, the frying process was simulated as 10 percent of control and 0.5 percent of moisture content of Gaussian white noise with 0 mean and unit variance.

With the final $\tilde{\gamma}$ of 0.04, the setpoint tracking responses and step disturbance rejection responses of the IMC controller based on the inverse of the modified Newton's method are displayed in Figures 6.9 and 6.10, respectively. The setpoint tracking responses show how the process outputs track the changes of the setpoint values. The step disturbance rejection responses show how the process outputs sustain the step disturbances that impact on the process outputs at certain instants during controller implementation.

Table 6.1 The Three Integral Error Objective Functions in the Tuning of $\tilde{\gamma}$ in an IMC Loop with the Process Model Inverse Using the Modified Newton's Method*

$\tilde{\gamma}$	ISE	IAE	ITAE
0.01	0.018381	0.059220	1.649493
0.02	0.010882	0.030475	0.515131
0.03	0.008664	0.020973	0.333839
0.04	0.007871	0.019237	0.316390
0.05	0.007797	0.020923	0.365465
0.06	0.008321	0.024489	0.449107
0.07	0.009652	0.030951	0.654441
0.08	0.012671	0.044000	1.230824

* From Huang et al. (1998a). With permission.

Table 6.2 The Three Integral Error Objective Functions in the Tuning of $\tilde{\gamma}$ in the IMC Loop with the Process Model Inverse Using the Gradient Descent Method*

$\tilde{\gamma}$	ISE	IAE	ITAE
0.001	0.026086	0.113393	0.855017
0.002	0.15497	0.077924	6.807420
0.003	0.012147	0.066104	6.449880
0.004	0.010648	0.060118	6.316670
0.005	0.009940	0.058497	6.308235
0.006	0.009683	0.059158	6.326198
0.007	0.009762	0.060914	6.372906
0.008	0.01158	0.063477	6.436001
0.009	0.010939	0.067334	6.529615
0.010	0.012320	0.073170	6.734011

* From Huang et al. (1998a). With permission.

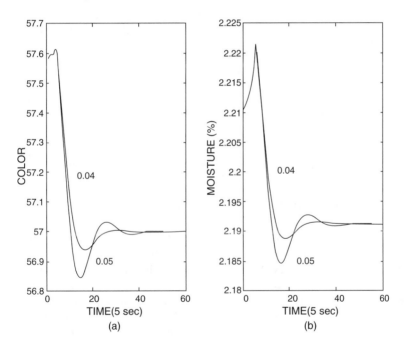

Figure 6.8 (a) IMC responses of color for $\tilde{\gamma} = 0.04$ and $\tilde{\gamma} = 0.05$ with the inverse of the modified Newton's method. (b) IMC responses of moisture content for $\tilde{\gamma} = 0.04$ and $\tilde{\gamma} = 0.05$ with the inverse of the modified Newton's method. (From Huang et al., 1998a. With permission.)

A similar approach was used to tune the 2 × 2 MIMO IMC controller based on the inverse with the gradient descent method.

Through the tuning shown in Table 6.2, the final $\tilde{\gamma}$ were 0.005 with IAE or ITAE and 0.006 with ISE. Figure 6.11 shows the output responses, $\tilde{\gamma} = 0.005$

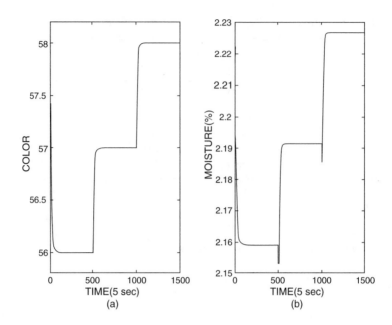

Figure 6.9 (a) IMC setpoint tracking response of color with the inverse of the modified Newton's method for $\tilde{\gamma} = 0.04$. (b) IMC setpoint tracking response of moisture content with the inverse of the modified Newton's method for $\tilde{\gamma} = 0.04$. (From Huang et al., 1998a. With permission.)

with IAE or ITAE was determined as the final value for smooth control based on the inverse of the gradient descent method.

With the final $\tilde{\gamma}$ value of 0.005, the setpoint tracking responses and the step disturbance rejection responses of the IMC controller are shown in Figures 6.12 and 6.13.

In this way a 2×2 MIMO neural network process one-step-ahead prediction model-based IMC loop has been established for quality control of a continuous snack food frying process. This control structure can compensate the time lags between process inputs and outputs by setting the prediction of the neural process model to track a feedback error signal in a closed loop. The controller was designed by computing each control action at each time instant using modified Newton's and gradient descent methods. Controller simulation results show that the established control loop can be tuned to deliver stable control of the product quality for the continuous snack food frying process.

6.3   *Predictive control*

The predictive control (PDC) is a different model-based control strategy from the IMC, which is based on process model multiple-step-ahead predictions. The basic idea of the PDC originated from dynamic matrix control (DMC) (Cutler and Ramaker, 1979). Figure 6.14 presents a schematic diagram

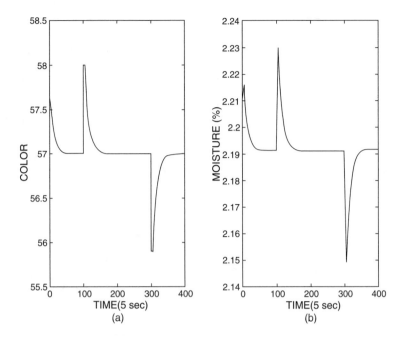

Figure 6.10 (a) IMC step disturbance rejection response of color in the inverse of the modified Newton's method with $\tilde{\gamma} = 0.04$. (b) IMC step disturbance rejection response of moisture content in the inverse of the modified Newton's method with $\tilde{\gamma} = 0.04$. (From Huang et al., 1998a. With permission.)

showing how the DMC works. In this figure, the optimizer is used to calculate the future input values. The optimization variables are $u(t + l - 1)$ ($l = 1, 2, \ldots, L$). They can be chosen to minimize an objective function in the form

$$J = \sum_{l=1}^{L} (y^s(t+l) - \hat{y}(t+l \,|\, t))^2 \qquad (6.37)$$

Typically, several future input values would be calculated, but only the first, $\hat{u}(t)$, is implemented as $u(t)$. One of the important features of this control strategy is that constraints on input and output can be incorporated into the formulation. This strategy constitutes a window moving horizon approach as shown in Figure 6.15. The center of the window is taken as the current time, t. Past and present values of input and output as well as future values of input are fed to the process predictor. The predictor outputs are the estimated process output values in the future. At the beginning of the implementation, the window is placed at the starting point. After each data presentation and calculation, the window moves Δt. This continues until the end of the implementation.

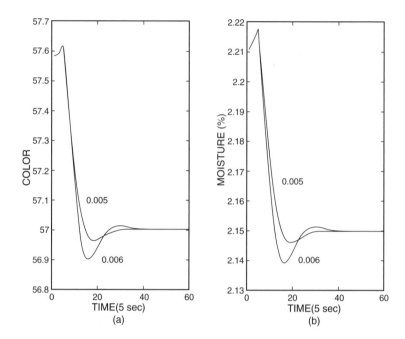

Figure 6.11 (a) IMC responses of color with $\tilde{\gamma} = 0.005$ and 0.006 in the inverse of the gradient descent method. (b) IMC responses of moisture content with $\tilde{\gamma} = 0.005$ and 0.006 in the inverse of the gradient descent method. (From Huang et al., 1998a. With permission.)

The preceding control strategy was extended to the generalized predictive control (GPC) (Clarke et al., 1987) based on a linear Controlled Autoregressive Integrated Moving-Average (CARIMA) model. The GPC has the following two important characteristics:

1. Assumption on the control signal, u—this control strategy is a receding-horizon method which depends on predicting the process's output over several steps based on an assumption about future control actions. This assumption is to define a control horizon, L_u, which is between the minimum objective horizon, L_1, and the maximum objective horizon, L_2, and, beyond it, all control actions are assumed to remain to be constant, or equivalently, all control increments are assumed to be zero.
2. Suppression on the control signal, u—in the GPC, the control signal, u, is incorporated in the objective function to be optimized

$$J = \sum_{l=L_1}^{L_2} (y^s(t+l) - \hat{y}(t+l \mid t))^2$$

$$+ \sum_{l=1}^{L_2} \lambda_l (\hat{u}(t+l-1) - \hat{u}(t+l-2))^2 \qquad (6.38)$$

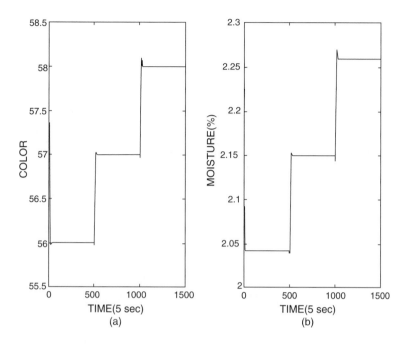

Figure 6.12 (a) IMC setpoint tracking response of color with $\tilde{\gamma} = 0.005$ in the inverse of the gradient descent method. (b) IMC setpoint tracking response of moisture content with $\tilde{\gamma} = 0.005$ in the inverse of the gradient descent method. (From Huang et al., 1998a. With permission.)

where $\{\lambda_i\}$ is the control weighting sequence. In this way, the control signal, u, is chosen according to the optimization problem and forced to achieve the desirable action.

It has been proven that the PDC has desirable stability properties for nonlinear systems (Keerthi and Gilbert, 1986; Mayne and Michalska, 1990). Further, it is possible to train a neural network to mimic the action of the optimization routine. This controller network is trained to produce the same control action as the optimization routine for a given process output. However, this training is usually not easy because it requires global invertibility of the process over the entire operating space. During training, the optimization routine can still be used to help achieve the desirable output (Hunt et al., 1992). The structure of this training is shown in Figure 6.16. Once the training is complete, the process model and optimization routine at the outer loop are no longer needed for implementation. This structure requires double calculations in the optimization routine and neural network training. Another approach to design the PDC controller is to use an on-line optimization routine directly to determine the future process inputs in terms of the minimization of the deviations between the desired and predicted process outputs over the multiple-step

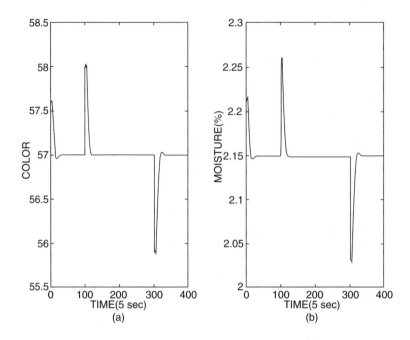

Figure 6.13 (a) IMC step disturbance rejection response of color with $\tilde{\gamma}$ = 0.005 in the inverse of the gradient descent method. (b) IMC set disturbance rejection response of moisture content with $\tilde{\gamma}$ = 0.005 in the inverse of the gradient descent method. (From Huang et al., 1998a. With permission.)

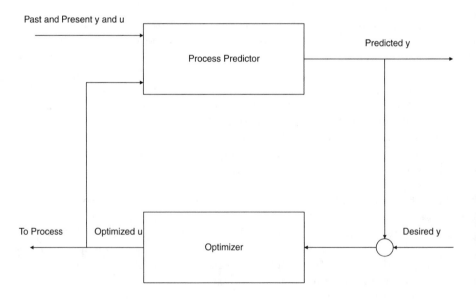

Figure 6.14 Structure of the DMC. (Adapted from Huang, 1995. With permission.)

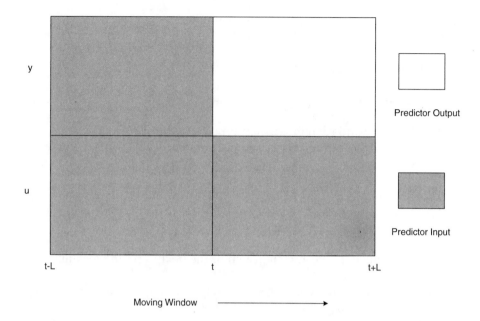

Figure 6.15 Scheme of a window moving horizon. (Adapted from Huang, 1995. With permission.)

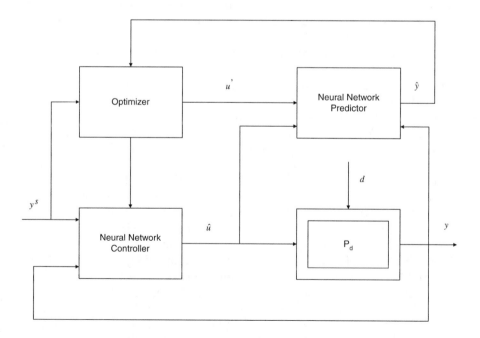

Figure 6.16 Structure of training for a neural network predictive controller. (Adapted from Huang, 1995. With permission.)

time horizon based on the neural process predictor. Thus, the neural network process model-based predictive controller for a process can be implemented as follows:

At each sampling instant,

1. Set up the future process output setpoint sequence, $y^s(t + l)$ and the control weighting sequence, λ_l.
2. Measure and save the current process output, $y(t)$.
3. Predict the future process output sequence, $\hat{y}(t + l|t)$, recursively, using the neural network process multiple-step-ahead predictor and formulate the objective function.
4. Minimize the objective function with an on-line optimization routine to give a suggested sequence of the future control actions, $u(t + l - 1)$.
5. Implement the first element of the control sequence, $u(t)$.
6. Shift the known values of y and u so that the calculations for the implementation can be repeated at the next sampling instant.

The on-line optimization routine plays an important role in PDC. Because of the nonlinearity of neural network process models, an analytical solution of the objective function generally cannot be obtained. If a numerical optimization method is used, the numerical solution to the objective function can be obtained iteratively. Because the optimization routine needs to be on-line, it is very important to consider the discontinuities owing to disturbances in the operation of the process in choosing a specific optimization method.

In general, there are two main types of optimization methods. One is to search the decreasing direction of the objective function based on the calculation of the gradient of the function. Examples are Least Square, Newton's method, and gradient descent. However, it is difficult to calculate gradients at different future instants at $t + l$ for the objective function designated for the neural network process model-based PDC because the detailed expansion is very complicated. Even though the gradients are calculated, the accumulation of the calculation error will have a significant impact on the calculation results when the calculation is of a high dimension. The objective function has a dimension of L_2, which could be high. The second type of optimization method only requires calculating the value of the objective function at different points instead of gradients to determine the decreasing direction of the objective function. This type of "direct" optimization method is adaptable to discontinuities in the process operation and avoids gradient calculation of a highly complicated and/or dimensional objective function. Among these direct optimization methods are simplex, random search, and conjugate direction. The method of conjugate direction is suitable for an objective function in quadratic form. This method can be used to set up an on-line optimization routine. The implementation of the neural network process model-based predictive controller using an on-line conjugate direction optimization routine is as follows.

The basic idea of the conjugate direction method is to search a specific direction through calculating the values of the objective function continuously until the algorithm converges to a minimum point. The optimization problem of the neural network process model-based PDC is to optimize a L_2 dimension nonlinear function. In order to minimize the objective function, $J(u(t), u(t+1), \ldots, u(t+L_2-1))$, L_2 future process inputs need to be determined. For this L_2 dimensional objective function, the algorithm for conjugate direction generation can be formulated as follows:

1. Set up the initial point, $\hat{\underline{u}}^0 = (\hat{u}^0(t), \hat{u}^0(t+1), \ldots, \hat{u}^0(t+L_2-1))^T$ and define the vector as

$$\underline{P}_i = \underline{e}_i, \quad i = 1, 2, \ldots, L_2$$

where e_i is the ith unit co-ordinate direction, $\underline{e}_i = (0, \ldots, 0, 1, 0, \ldots, 0)^T$, here the ith component is 1, and the rest are 0.

2. Set $i = 1$, calculate ρ_i to minimize $J(\hat{\underline{u}}^{i-1} + \rho_i \underline{P}_i)$, and define

$$\hat{\underline{u}}^i = \hat{\underline{u}}^{i-1} + \rho_i \underline{P}_i$$

and then set $i = i + 1$, and repeat this until $i = L_2$.

3. Set $\underline{P}_i = \underline{P}_{i+1}$ ($i = 1, 2, \ldots, L_2 - 1$), $\underline{P}_{L_2} = \hat{\underline{u}}^{L_2} - \hat{\underline{u}}^0$.

4. Calculate ρ_{L_2} to minimize $J(\hat{\underline{u}}^{L_2} + \rho_{L_2} \underline{P}_{L_2})$, and define $\hat{\underline{u}}^0 = \hat{\underline{u}}^{L_2} + \rho_{L_2} \underline{P}_{L_2}$, then proceed to Step 2. After the preceding calculations are done L_2 times, continue to Step 1 until $\|\rho_{L_2} \underline{P}_{L_2}\| < \varepsilon$ where ε is the predefined error for termination of the computation. At this moment, $\hat{\underline{u}}^0$ is the minimum point, and $J(\hat{\underline{u}}^0)$ is the minimum value of the objective function.

As previously mentioned, if there are constraints on the process inputs or outputs, these constraints can be incorporated into the optimization problem. It is obvious that optimization is more difficult to perform with constraints than without them. Numerical optimization methods without constraints are better established than those with constraints. Therefore, it is desirable to transform the problem of optimization with constraints into a problem without constraints, solve the optimization problem without constraints, and use the solution as the approximation to the optimization problem with constraints. For convenience in algorithm development and implementation, it is better to use the same numerical optimization methods to solve the optimization problem with and without constraints. This transformation can be done by establishing a new objective function

$$\tilde{J}(\hat{u}, r) = J(\hat{u}) + \sum_{i=1}^{m} r_i \psi_i^2(\hat{u})$$

where $J(\hat{u})$ is the objective function of the optimization problem, m is the number of the process constraints, r_i is the ith penalty constant, and $\psi_i(\hat{u})$

is the *i*th process constraints on inputs and outputs in which there are no outputs explicitly because the process outputs are functions of process inputs. The second term of the preceding equation is called the penalty function. It can be understood as a certain cost needed in the objective function because of constraint violation in the control action, $\hat{u}$.

Intuitively, the larger the values of *r*, the closer the solution is to the original problem without constraints. However, the larger the *r* values, the larger the cost. Consequently, the determination of reasonable *r* values is important in practical optimization. If *r* values begin from small values, increase gradually, and the known solution is used to initialize the next optimization, the algorithm may converge to the solution of the minimization with constraints. This produces a sequence of *r* values and a sequence of optimization problems corresponding to these different *r* values while each of these optimization problems still can be solved through the conjugate direction method.

Fuzzy logic-based control has emerged as a promising approach for complex and/or ill-defined process control. It is a control method based on fuzzy logic (Zadeh, 1965). However, formal methods to identify fuzzy inference rules do not exist. Several suggested approaches lack adaptability or a learning algorithm to tune the membership functions in fuzzy logic (Jang, 1992). Control engineers have studied a self-learning fuzzy controller since Procyk and Mamdani (1979) developed it.

A self-learning fuzzy controller with an ANN estimator was designed for predictive process control based on Choi et al.'s work in 1996.

The design of the fuzzy controller is composed of a comparator, a fuzzy controller, and an ANN estimator as shown in Figure 6.17. The ANN estimator predicts process output vector, $\hat{y}(t+1)$, at the time, *t*, based on the extension of Eq. (5.9)

$$\hat{\underline{y}}(t+1) = \hat{f}(\hat{\underline{y}}(t), \hat{\underline{y}}(t-1), \ldots, \hat{\underline{y}}(t-p+1), \underline{u}(t),$$

$$\underline{u}(t-1), \ldots, \underline{u}(t-q+1), \hat{\Theta}) \tag{6.39}$$

Several fuzzy inference systems using ANNs have been proposed for automatic extraction of fuzzy rules and tuning of fuzzy membership functions (Jang, 1992; Horikawa et al., 1992). These approaches realized the process of fuzzy reasoning by the structure of an ANN and express the parameters in fuzzy reasoning by the connection weights of an ANN. This methodology automatically identifies the fuzzy rules and adaptively tunes membership functions by modifying the connection weights of the networks through a BP training algorithm. This kind of fuzzy controller is often called a fuzzy neural network (Horikawa et al., 1992) or an adaptive network-based fuzzy controller (Jang, 1992).

Figure 6.18 shows a design of a simplified neuro-fuzzy 2 × 2 controller network. The main features of the controller are

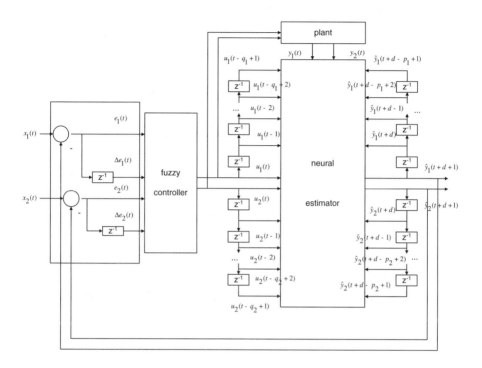

Figure 6.17 Block diagram of a neuro-fuzzy controller system. (Adapted from Choi et al., 1996. With permission.)

1. The gradient descent-based BP is applied to find a set of controller network weights that minimize the objective function, J, over all time steps

$$J = E\left(\sum_{i=1}^{2}\sum_{k=1}^{T}\alpha_i\left\|y_i^d - y_i^k\right\|^2 + \sum_{i=1}^{2}\sum_{k=0}^{T-1}\beta_i\left\|u_i^k\right\|^2\right) \qquad (6.40)$$

    where E is the error measure, y_i^d is the ith desired process output, y_i^k is the ith actual process output, u_i^k is the ith controller's output at the kth time step, α_i is the coefficient that is the importance factor of each error component, β_i is the coefficient which determines the rate of control energy minimization, and T represents training time steps.

2. Fuzzy membership functions are composed of sigmoid functions with parameters (a_i, b_i) in the premise part of each fuzzy rule, where a_i determines the gradients of a sigmoid function and b_i determines the central position of a sigmoid function. A scaling parameter set, s, scales the input to the range of –1 to +1. Any other continuous and piecewise dissimilar functions, such as trapezoidal, bell-shaped, or triangular-shaped membership functions, can also be used. $\mu_{A_i}(x)$

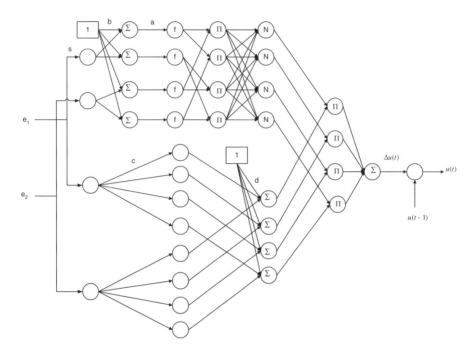

Figure 6.18 A simplified neuro-fuzzy inference network. (Adapted from Choi et al.,1996. With permission.)

determines the degree to which the input, x, satisfies the quantifier A_i:

$$\mu_{A_i}(x) = \frac{1}{1 + \exp[-a_i(sx + b_i)]} \tag{6.41}$$

3. The multiplication operator labeled Π, is used for the calculation of the firing strength, ω_i, of a rule

$$\omega_i = \mu_{A_i}(e_1(t)) \times \mu_{B_i}(e_2(t)) \times \mu_{C_i}(e_2(t-1)) \times \mu_{D_i}(e_2(t-1)) \quad (i = 1, 2,\ldots, p) \tag{6.42}$$

where A_i, B_i, C_i, and D_i are fuzzy membership functions for each input.

4. A fuzzy if–then rule (Takagi and Sugeno, 1983) is used for fuzzy reasoning. The output of each rule is a linear combination of input variables and a constant term. The final output of the controller is the weighted average of each rule's output. The rule can be stated as

If $e_1(t)$ is A_i and $e_2(t)$ is B_i, and $e_1(t-1)$ is C_i, and $e_2(t-1)$ is D_i, then $u = g[\omega_i, f_i]$ $(i = 1, 2,\ldots, p)$

$$u = \sum_{i=1}^{p} \hat{\omega}_i f_i(e_1(t), e_2(t), e_1(t-1), e_2(t-1)) \tag{6.43}$$

where

$$f_i(e_1(t), e_2(t), e_1(t-1), e_2(t-1))$$
$$= d_i + c_{i1}e_1(t) + c_{i2}e_2(t) + c_{i3}e_1(t-1) + c_{i4}e_2(t-1) \qquad (6.44)$$

and

$$\hat{\omega}_i = \frac{\omega_i}{\sum_{i=1}^{p} \omega_i} \qquad (6.45)$$

where c_{ij} ($j = 1, 2, 3, 4$) and d_i are weight parameters in the consequence part of each fuzzy rule, and ω_i is the ratio of the ith rule's firing strength to the sum of all rules' firing strengths.

5. The controller has four inputs of two errors and two error changes. The errors are the differences between the target values and the actual values of process outputs. The error changes are the differences between the current error at time, t, and the previous errors at time, $t - 1$. Also, the controller determines the changes of the process input vector, $\Delta \underline{u}(t)$. The final output vector of the controller, $\underline{u}(t)$, becomes $\underline{u}(t - 1) + \Delta \underline{u}(t)$.

6. The strict gradient descent method for the controller network training is used. The premise parameters a_i, b_i, and the consequent parameter c_{ij}, and d_i of the designed controller are updated to identify fuzzy rules and to tune fuzzy membership functions as the training proceeds to minimize the objective function in Eq. (6.40). The parameter update equation is

$$w_{ji}^{(k)}(t+1) = w_{ji}^{(k)}(t) + \eta \delta_j^{(k)} O_i^{(k-1)} \qquad (6.46)$$

where in the output layer

$$\delta_j^{(k)} = (y_j^d - O_j^{(k)}) f'(i_j^{(k)}) \qquad (6.47)$$

and in the hidden layer

$$\delta_j^{(k)} = f'(i_j^{(k)}) \sum_l \delta_l^{(k+1)} w_{lj}^{(k+1)} \qquad (6.48)$$

where $O_j^{(k-1)}$ is the output of the jth unit in the $(k-1)$th layer and $i^{(k)}$ is the input of the jth unit in the kth layer.

The error term located in the front layer of the multiplication operation, Π, is expressed as

$$d_j^{(k)} = f^1[i_j^{(k)}]\left\{\sum_l \delta_l^{(k+1)} w_{lj}^{(k+1)}\left[\prod_{i\neq j} w_{lj}^{(k+1)} O_i^{(k)}\right]\right\}$$ (6.49)

There are two kinds of output functions in this controller network—a linear output function and a sigmoid output function. The derivative of linear output function is $f' = 1$, and the derivative of sigmoid output function is $f' = f(1 - f)$. The unmarked node is merely to distribute inputs to the next layer in Figure 6.18. The proposed controller's structure can be easily extended to the general MIMO case. For multiple inputs, the same structures in Figure 4.19 can be added parallel to the structure for a single input.

6.3.1   Example: Neuro-fuzzy PDC for snack food frying process

Choi et al. (1996) utilized the neuro-fuzzy control structure described previously to perform PDC of the snack food frying process. For the snack food frying process, a significant feature is the long time lag between the inputs and outputs of the process. The representation of the ANN estimator given in Eq. (6.39) should be extended into the following equation to express the time lag, d, explicitly,

$$\hat{y}(t+d+1) = \hat{f}(\hat{\underline{y}}(t+d), \hat{y}(t+d-1),\ldots, \hat{y}(t+d-p+1),$$

$$\underline{u}(t), \underline{u}(t-1),\ldots, \underline{u}(t-q+1), \hat{\Theta})$$ (6.50)

The ANN estimator predicts process output vector, $\hat{\mathbf{y}}(t + d + 1)$, at time, t. The ANN estimator is a time delay multiplayer perceptron with output feedback with the same structure of ERECN described in Chapter 5. The training, testing, and validation of the ANN estimator are similar to the description in Section 5.3, and the results are shown in Section 5.3.1. This example will focus on the work on controller development.

The neuro-fuzzy predictive controller was applied to the snack food frying process by computer simulation. Before training the controller, the process estimator was trained for modeling the frying process. Then, the controller was trained with a training data set consisting of initial process states and the desired outputs on color and moisture contents. At time $t = 0$, the controller generates a process input and the process estimator predicts the process output based on the controller's output at a later time $t = r$ (r is the sampling time). The process estimator repeats the prediction until the process output at time $t + d$, the actual process output resulting from controller's output at time t and the previous inputs to the process up to time t are obtained. The controller determines its next output according to the error between the reference input and the predicted process output for the training time steps. Controller weights are updated once according to an accumulated error over all training time steps. This procedure is continued until the objective function is minimized.

The training data sets for the controller training were selected from the experimental data, also used for training the process estimator. The controller was trained with training data that described the desired response including the transient response; however, it is very difficult to predefine the best-desired response in cases in which the actual response cannot be predicted. Also, there is no guarantee of the controller's training capability for specific training data.

The goal was to get both the final desired output and an acceptable transient response. In this study, the batch learning method was used to minimize error over all time steps, and the number of training time steps was selected to be 200.

After training the controller, the desired process output was obtained within 100 iterations. As the training proceeded further, the MSE decreased. Because it was observed that excessive overshoot in the transient response appeared when the controller was overtrained, it appeared that memorization could negatively affect the desired control performance. This characteristic may be a result of the batch learning of the controller.

Only two training data sets were used for the controller training. The initial conditions were (30.0, 28.5; 7.2, 6.6) and (27.3, 28.5; 6.0, 6.6). First and third values in the parentheses represented initial states which were close to boundary conditions of color and moisture content, respectively. For instance, the color of the product was too light at 30.0 and it was too dark at 27.3. Second and fourth values represented the desired values of color and moisture content, respectively.

The numeric values used to express conditions were not real values but rescaled values. In the controller training, the initial values of parameters c and d were set to zero, and the Least Mean Squares was applied once at the very beginning to get the initial values of the consequent parameters before the gradient descent took over to update all parameters.

Eq. (6.40) contains two kinds of coefficients, α and β, in the objective function. The parameter α weights the importance of color and moisture content. It is impossible to get both the desired color and the desired moisture content with any values of α because the color and the moisture content are not controllable separately because they are chemically related to each other. The color was considered the most important quality in this case. If the controller's output is not used as part of the objective function, the controller outputs are usually large. Control energy can be minimized by making the controller learn to make u^k smaller in magnitude. Figure 6.19 shows the effect of the control energy coefficient of exposure temperature, β_1, on color, moisture content, and exposure temperature. Dashed, solid, and

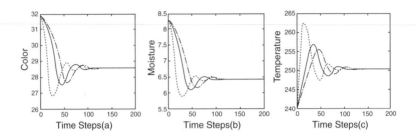

Figure 6.19 Effect of control coefficient, β_1 (dotted, solid, and dash–dot lines corresponding to $\beta_1 = 0.002$, $\beta_1 = 0.005$, $\beta_1 = 0.007$, respectively). (From Choi et al., 1996. With permission.)

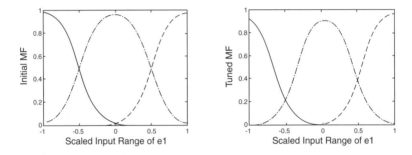

Figure 6.20 Typical example of a membership function tuning. (From Choi et al., 1996. With permission.)

dash–dot lines correspond to β_1 equal to 0.002, 0.005, and 0.007, respectively. It was observed that as β_1 became larger, a shorter settling time could be obtained; however, the controller outputs were large. The most acceptable result was obtained for $\beta_1 = 0.005$ because it resulted in a small overshoot in the transient response. The three membership functions were used for every input in this study. Figure 6.20 shows the initial membership functions and trained membership function of the error of exposure temperature as an example.

Figure 6.21 demonstrates the effect of the control energy coefficient of exposure time, β_2. Solid, dotted, and dash–dot lines correspond to $\beta_2 = 30$, $\beta_2 = 10$, and $\beta_2 = 0$, respectively. As β_2 increased, better transient responses of color and moisture content could be obtained, and the change of exposure temperature was small.

Untrained initial conditions were given in order to test the adaptation of the controller. Figure 6.22 shows the results where dashed, solid, dash–dot, and dotted lines correspond to the initial conditions, Case 1 (30.0, 28.5; 7.2, 6.6), Case 2 (27.3, 28.5; 6.0, 6.6), Case 3 (32.0, 28.5; 8.1, 6.6), and Case 4 (26.7, 28.5; 6.0, 6.6), respectively. Cases 1 and 2 are the conditions used in the controller training, shown again for comparison. Case 3 has larger differences between the initial value and the desired value of color or moisture

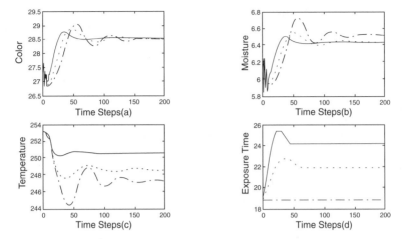

Figure 6.21 Effect of the control coefficient, β_2 (dotted, solid, and dash–dot lines corresponding to $\beta_2 = 0$, and $\beta_2 = 10$, and $\beta_2 = 30$, respectively). (From Choi et al., 1996. With permission.)

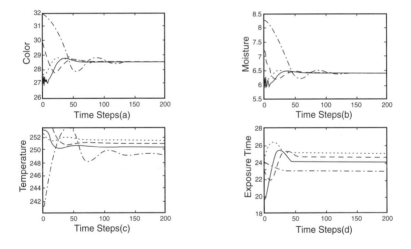

Figure 6.22 The adaptation of the controller for untrained conditions. (From Choi et al., 1996. With permission.)

content than do Cases 1 and 2. The initial states of color and moisture content are almost the same in both Case 2 and 4. However, the responses are different from each other because they have different previous values of exposure time at the initial state used as the inputs to the controller. The settling time of Case 3 was much larger than other cases when the trained controller was used. To shorten settling time in the cases having a large initial difference, such as Case 3, the controller output for exposure temperature was increased in proportion to the initial difference of color between the test data set and the training data set. The control task was performed success-fully in all initial conditions as shown in Figure 6.22. This result confirms that the controller has an adaptive capability for untrained initial conditions.

References

Choi, Y. S., Whittaker, A. D., and Bullock, D. C., Predictive neuro-fuzzy controller for multivariate process control, *Trans. ASAE*, 39(4), 1535, 1996.

Clarke, D. W., Mohtadi, C., and Tuffs, P. S., Generalized predictive control Part I. The basic algorithm, *Automatica*, 23(2), 137, 1987.

Cutler, C. R. and Ramaker, B. L., Dynamic matrix control—a computer control algorithm, AIChE 86th National Meeting, 1979.

Economou, C. G., Morari, M., and Palsson, B. O., Internal model control. 5. Extension to nonlinear systems, *Ind. Engin. Chem. Process Design Devel.*, 25, 403, 1986.

Garcia, C. E. and Morari, M., Internal model control. 1. A unifying review and some results, *Ind. Engin. Chem. Process Design Devel.*, 21, 308, 1982.

Horikawa, S., Furuhashi, T., and Uchikawa, Y., On fuzzy modeling using fuzzy neural networks with the backpropagation algorithm, *IEEE Trans. Neural Net.*, 3(5), 801, 1992.

Huang, Y., Snack food frying process input–output modeling and control through artificial neural networks, Ph.D. dissertation, Texas A&M University, College Station, TX, 1995.

Huang, Y., Whittaker, A. D., and Lacey, R. E., Neural network prediction modeling for a continuous, snack food frying process, *Trans. ASAE*, 41(5), 1511, 1998.

Huang, Y., Whittaker, A. D., and Lacey, R. E., Internal model control for a continuous, snack food frying process using neural networks, *Trans. ASAE*, 41(5), 1519, 1998a.

Hunt, K. J., Sbarbaro, D., Zbikowski, R., and Gawthrop, P. J., Neural networks for control systems—a survey, *Automatica*, 28(6), 1083, 1992.

Jang, J. R., Self-learning fuzzy controllers based on temporal back propagation, *IEEE Trans. Neural Net.*, 3(5), 714, 1992.

Keerthi, S. S. and Gilbert, E. G., Moving-horizon approximation for a general class of optimal nonlinear infinite-horizon discrete-time systems, in *Proceedings of the 20th Annual Conference on Information Science and Systems*, Princeton University, 1986, 301.

Mayne, D. Q. and Michalska, H., Receding horizon control of nonlinear systems, *IEEE Trans. Autom. Contr.*, 35, 814, 1990.

Prett, D. M. and Garcia, C. E., *Fundamental Process Control*, Butterworths, Boston, 1988.

Procyk, T. J. and Mamdani, E. H., A linguistic self-organizing process controller, *Automatica*, 15(1), 15, 1979.

Scales, L. E., *Introduction to Nonlinear Optimization*, Springer-Verlag, New York, 1985.

Takagi, T. and Sugeno, M., Derivation of fuzzy control rules from human operator's control actions, in *Proc. IFAC Symp. Fuzzy Information, Knowledge Representation and Decision Analysis*, Marseille, 1983, 55.

Zadeh, L. A., Fuzzy sets, *Inform. Contr.*, 8(3), 338, 1965.

chapter seven

Systems integration

In the previous chapters, the components of systems for food quality quantization and process control were discussed and described in detail in the chain of data acquisition, data processing and analysis, modeling, classification and prediction, and control. This concluding chapter discusses the principles of systems integration for food quality quantization and process control. The techniques of systems development, especially software development, for food quality quantization and process control are discussed. Through these works, we hope that readers can visualize the processes of food quality quantization and process control systems integration and development.

7.1 Food quality quantization systems integration

As described in previous chapters, systems of food quality quantization consist of data sampling, data measurement and collection, data processing and analysis, modeling, and classification and prediction. The integration of these components into a system needs hardware and software development. The development of hardware in food quality quantization is typically for data measurement and collection while the development of software occurs throughout the complete process of food quality quantization. Computers are in command of the process of automated food quality evaluation. Figure 7.1 shows the structure of the integration for food quality quantization systems.

In the structure, food samples are prepared based on the experimental design. With the samples, data of x and y are measured and collected. As described in previous chapters, x usually represents electronic scans to produce one-dimensional signals or two-dimensional images that can be measured and collected by a DAQ system. y usually represents physical properties of food samples that can be measured and recorded. After data are collected, the data are preprocessed, and features are extracted from the signal x^p as the input $\tilde{x}$ to the process of food quality quantization. Then, the cross-correlation analysis is performed between $\tilde{x}$ and $\tilde{y}$. Next, $(\tilde{x}, \tilde{y})$ is used to model the relationship between $\tilde{x}$ and $\tilde{y}$ for the purpose of food

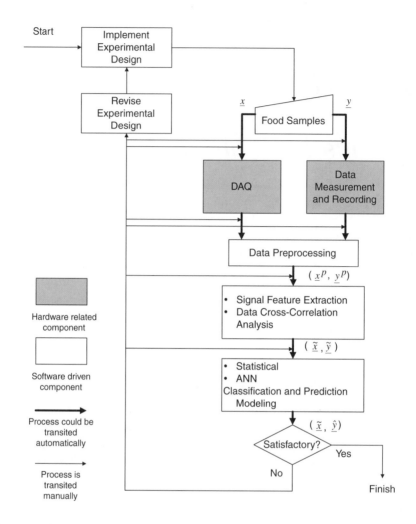

Figure 7.1 Structure of systems integration for food quality quantization.

sample classification or food attribute prediction with the statistical and/or ANN methods. If the result of the classification or prediction, $\hat{y}$, is satisfactory, the procedure stops with a final model for food quality quantization. Otherwise, the procedure needs to be repeated to make adjustments in data collection, data preprocessing, data analysis, or modeling, or even to revise the previous experiment design and run the whole process again.

In the process of food quality quantization, the transitions inside the system from data measurement, data preprocessing, data analysis, and modeling up to classification and prediction can be automatic. As long as the protocols for data representation are consistent between components, the communication between different components should not be a problem.

The components of DAQ for $\underline{x}$ and data measurement and recording for $\underline{y}$ need electronic hardware support (sensors, signal conditioning, A/D and D/A conversions, etc.) with the communication of a console computer (usually PC). The rest of the components are all software driven. The computation can be done in the same computer or a separate, more powerful one.

7.2   Food quality process control systems integration

The integration of components such as data sampling, data acquisition, data processing and analysis, modeling, prediction, and control into a system needs hardware and software development. For example, hardware development is needed for turning on actuators based on signal generation from the implementation of experiment design and data acquisition. Software development occurs throughout the complete process of food quality process modeling and control. Figure 7.2 shows the structure of the integration for food quality process control systems.

In the structure, the food process is perturbed by implementing experimental designs to generate input signals, $\underline{u}(t)$, for turning on the actuators on the line while logging the output signal, $\underline{y}(t)$. Unlike food quality quantization systems that are static in general, food quality process control systems are dynamic. The DAQ system measures and collects the input signal, $\underline{u}(t)$, and the dynamic response, $\underline{y}(t)$, of the process. Then, the data are preprocessed into $(\underline{u}^p(t), \underline{y}^p(t))$. Auto- and cross-correlation analyses are performed on $(\underline{u}^p(t), \underline{y}^p(t))$ to produce a compact data set $(\underline{\tilde{u}}(t), \underline{\tilde{y}}(t))$. The compact data set is used to model the relationship between the process inputs and outputs. The modeling focuses on one-step-ahead prediction, $\hat{y}(t+1|t)$, or multiple-step-ahead prediction, $\hat{y}(t+l|t)$ $(l > 1)$, using statistical or ANN methods, depending on the degree of nonlinearity of the process dynamics. When the process model is ready, the controller can be designed based on inverse dynamics to achieve the specified process output setpoint, $\underline{y}^s(t)$. The result of the controller design, $\underline{\hat{u}}(t)$, is simulated. If it is satisfactory, the system plugs $\underline{\hat{u}}(t)$ into the experimental design to implement the control system; otherwise, the system may need adjustments in the components of data preprocessing, data correlation analysis, process prediction modeling, or controller design. During the implementation of the system, a component evaluates the performance of the control system based on certain criteria of minimization of the mismatch between $\underline{\tilde{y}}(t)$ and $\underline{y}^s(t)$. If the performance evaluation proves satisfactory, the control system can accept the design and implement it in the real line; otherwise, the system needs to go back to revise the experimental design.

In the process of food quality process control, the transitions in the system beginning with signal generation, and proceeding to data acquisition, data preprocessing, data analysis, modeling, and control can be automatic. As long as the protocols for data representation are consistent between

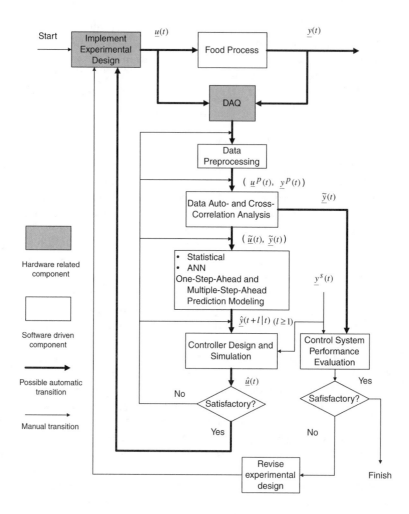

Figure 7.2 Structure of systems integration for food quality process control.

components, the communication between different components should be no
problem.

The components for turning on actuators for generated signal and DAQ
for data measurement and recording for $\underline{u}(t)$ and $\underline{y}(t)$ need electronic hardware support (sensors, signal conditioning, A/D and D/A conversions, etc.)
with the communication of a console computer (usually PC). The rest of the
components are all software driven. The computation can be done in the
same computer or a separate more powerful one.

For control systems, the modeling and control computations can be
done off-line with a supervisory computer. For example, the training of
ANN process model and controller (inverse of the forward model) can be
performed off-line because ANN training is time-consuming. Once the
training is done, the model and controller can be plugged into the system

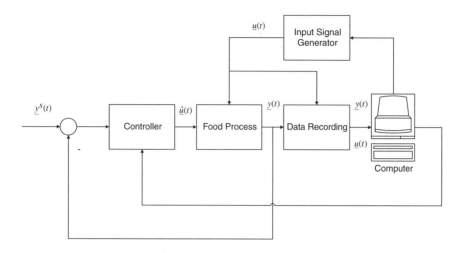

Figure 7.3 Structure of off-line modeling and control computing.

for implementation. The structure of the off-line modeling and control computing is shown in Figure 7.3. Either of the computations can be done on-line with a process computer. For on-line computation, the regular algorithms, which are usually in batch form, need to be transformed into recursive form. For example, in Chapter 4 the batch algorithm of least squares was given. However, because this algorithm needs enormous space in computer memory, it cannot be used in on-line computation. The problems can be transformed into a recursive algorithm for solving. The basic idea of recursive computing can be explained with the following formula

$$\text{new estimate } \hat{\beta}(k) = \text{old estimate } \hat{\beta}(k-1) + \text{modification}$$

In this formula, the new estimate $\hat{\beta}(k)$ is obtained with the modification on the basis of the old estimate $\hat{\beta}(k-1)$. Therefore, the computation needs significantly less computer memory, and it can be used in on-line computation. Figure 7.4 shows the structure of on-line modeling and control computing.

In control systems, traditionally there are electromechanical devices such as switches and relays used to control the operation of a plant and machinery. Such systems are flexible in design and easy to understand. However, when they are used to realize a complex control "logic," for example, a switch may not directly turn a device on while the switch may have different impacts on the system to initialize a complex logic, the circuit needs to be redesigned and rewired. Programmable logic controllers (PLCs) have been used in industry since late 1960 and they have been established as powerful tools in controlling

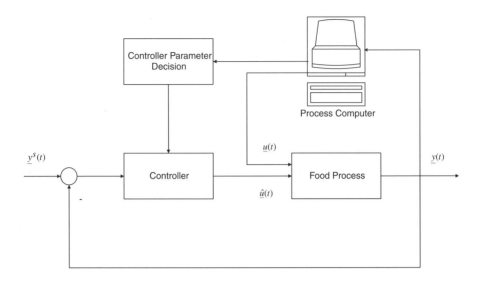

Figure 7.4 Structure of on-line modeling and control computing.

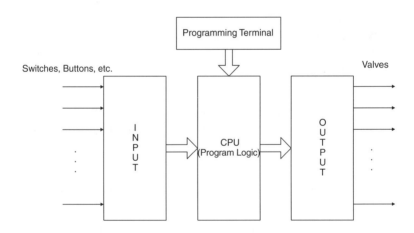

Figure 7.5 Structure diagram of a PLC.

the operation of a plant and machinery. A PLC is a device that can be programmed to perform a control function. With a PLC, virtually any complex control logic can be realized without redesigning and rewiring the circuit as long as the connecting ports of the switches to the input of the PLC and the devices in the output of the PLC are known to the programmer. Figure 7.5 shows the structure diagram of a PLC. From the diagram, it can be seen that the input and output modules can be configured to provide different numbers of inputs and outputs, and the relationship between them can be programmed through the programming terminal of the PLC to the CPU (central processing unit).

PLCs have a strong ability to perform mathematical functions. They can be used in process control. For example, in quality process control of the continuous snack food frying process, the oil temperature, product conveyor motor speed, and so on, can be controlled using PLCs. Although we do not discuss PLCs in detail here, we mentioned them to focus your attention on the technology. PLCs are very useful in practical process control. Interested readers can refer to the papers and books on this topic for details.

7.3 Food quality quantization and process control systems development

Similar to general systems development, the development of systems for food quality quantization and process control should observe the principle of systems development life cycle. In general, the process of systems development consists of five phases: systems analysis, systems design, systems development and implementation, systems operation and maintenance, and systems evaluation. Systems analysis includes user requirement analysis, system economic and technical feasibility analysis, and system logical modeling. Systems design establishes a system physical model. Systems development and implementation is for detailed programming and system testing. Systems operation and maintenance is for system application. Then, if the evaluation shows the system cannot provide sufficient benefit, the system may need to revert to systems analysis to initialize a new process of systems development. That is what the life cycle implies. Figure 7.6 shows the flow of the systems development life cycle. Actually, Systems Development Life Cycle is a process by which systems analysts, software engineers, programmers, and end-users work together to build information systems and computer applications. It is a project management tool used to plan, execute, and control systems development projects. This is useful for food quality quantization and process control systems development, especially, for complex and large computer modeling and control systems. Interested readers can refer to the papers and books on this topic.

As discussed in the previous two sections, software development is essential in the process of food quality quantization and process control systems development. Software development is involved in data acquisition, data analysis, modeling, classification and prediction, and control.

Programming is the work to realize software development using programming languages. Different computer programming languages, from machine language and assembly language to high-level language, are used for different purposes in different situations. Machine language is the native language of a computer. Machine language is pure binary code and is difficult for laymen to understand. To help people manage the code easier, a direct translation, that is, command by command, of the binary code to a symbolic form can be made. This is the function of assembly language. With the help of the assembler, people can better understand the

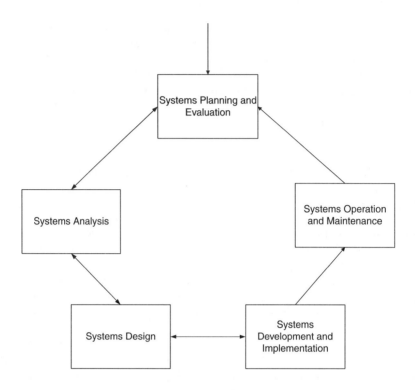

Figure 7.6 Flow of a systems development life cycle.

operation of their computers and make faster and smaller capable programs using a language made of symbolic representations. High-level language provides a higher level of abstraction than assembly language. Usually one high-level language command represents one or a number of machine or assembly commands. FORTRAN, Pascal, and C are representative high-level languages. FORTRAN is for scientific computing. Pascal is a structural programming language. C is a high-level programming language developed by Bell Labs (Murray Hill, NJ) in the mid of 1970s (Ritchie and Kernighan, 1978). Originally, C language was closely related to the UNIX operation system (Bourne, 1982), an innovation of Bell Labs. Now, it has become a popular programming language independent of UNIX. Actually, because C language was originally designed to write systems programs, it is much closer to assembly language than most other high-level languages. Sometimes, C is considered a language between assembly and high-level language. These characteristics allow C language to write very efficient code.

In food process DAQ systems, C language can be used to write programs to communicate between sensors, the DAQ board, and the computer. C language system provides a number of function prototypes for this. Consequently, C language is a language people need to consider when developing food quality quantization and process control systems.

In discussions about programming for DAQ systems, LabVIEW, a graphical programming development environment developed by National Instruments Corporation (Austin, TX), is worth mentioning. Unlike other text-based programming languages, LabVIEW uses a graphical programming language, G, to create programs in block diagram form. LabVIEW can be used for data acquisition and control, data analysis, and data presentation. LabVIEW has been used in food quality DAQ systems development. For example, the electronic nose system and the meat elastography systems described in previous chapters used LabVIEW in DAQ systems programming.

In algorithmic computing for food process data analysis, modeling, classification and prediction, and control, the programs can be coded using languages like FORTRAN and C. MATLAB, a matrix-based integrated technical computing environment developed by The MathWorks, Inc. (Natick, MA) is a powerful tool in computing. MATLAB combines numeric computation, advanced graphic and visualization, and a high-level programming language. Besides basic commands and functions, MATLAB provides various application-specific toolboxes. The toolboxes we used for food quality quantization and process control are statistics, signal processing, system identification, image processing, neural networks, wavelet analysis, and so on.

The emergence and development of object-oriented programming (OOP) have revolutionized the way software engineers design systems and programmers write programs. In OOP, programmers can define both the type of a data structure and the types of operations with the data structure. Now, the data structure is an object that includes both data and functions. OOP is characterized by concepts such as inheritance, dynamic binding, polymorphism, and information hiding. OOP is a tool to enhance the efficiency and reliability of code and provides an effective way for extension and reuse of code. In systems development for food quality quantization and process control, OOP can be used to structure the basic data representations as objects, such as vector, matrix, and so on, and design the classes for different methods for statistical data analysis, modeling, classification and prediction, and control. For example, for ANN training, under the general ANN training class, there are different supervised and unsupervised training algorithms. These can be designed as classes in the sense of OOP. Figure 7.7 shows the class hierarchy. In the hierarchy, each child class inherits everything from a parent class. In developing object-oriented software, it is important to have this kind of class hierarchy diagram, which clarifies the relationships between classes and helps extend the hierarchy when necessary.

There are several programming languages available to support OOP. Smalltalk, C++, and Java are the most popular ones. Smalltalk is a pure object-oriented language. C++ is actually an extension and improvement from C language. C++ also was developed by Bell Labs (Stroustrup, 1985). C++ adds object-oriented features to C, and it is a powerful programming language for graphical applications in Windows environments. Java is another object-oriented language similar to C++, but it is simplified to eliminate language features that cause common programming errors. Java interpreters execute

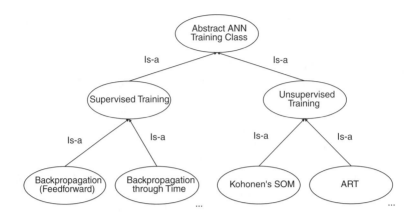

Figure 7.7 Class hierarchy for ANN training algorithms.

bytecode, compiled from Java source code. Because Java virtual machines (JVM), that is, Java interpreters and runtime environments, exist in most operating systems or platforms, compiled Java code can run on most computers. Java is platform independent and can be used to develop either stand-alone or Web-based applications.

Information access sites or World Wide Web home pages have been developed using the information superhighway or Internet in various areas, including food science and engineering. Today, millions of people, worldwide, access information on various topics through the Internet. Through a home page on the Internet, millions of people could potentially access information as the clients, and the clients could have access to a host of information providers. Food information systems can be integrated, managed, and used through the Internet for the food information needs of our clients.

Letting databases work on the Web is a promising alternative to developing and applying the information in the databases. Linking the Web pages to databases instantly adds interactivity, functionality, and excitement. Web database provides a database with dynamically distributed interfaces for better information sharing (Huang et al., 1999). Figure 7.8 shows the structural diagram of the Web database. In the diagram, the functions can be information retrieving for information management, for example, information retrieving for faculty, facilities, and resources for an institute of food science and engineering. The functions can be information processing for system integration, for example, information processing for data analysis, modeling, classification and prediction, and control for food process automation. All of the functions are co-ordinated by the implementation of system management programs through a database on the request from a Web server. The Web database is driven by the client's request to the Web page through the Web server.

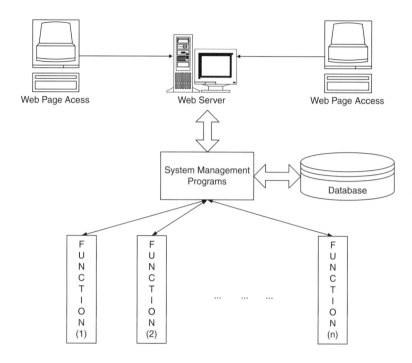

Figure 7.8 Structural diagram of a Web database.

7.4   Concluding remarks

We have examined all of the components, including data acquisition, data analysis, modeling, classification and prediction, and control used for building complete systems for food quality quantization or process control. The development of electronics and computer science provides a solid foundation for developing high-performance food quality quantization and process control systems. The performance of food quality quantization and process control systems will increase with the development of electronics and computer science. Of course, careful consideration must be given to the use of new hardware and software technologies. We need to perform technical and economic feasibility studies based on the conditions of specific food objects or processes.

We need to test the techniques, methods, and devices on a limited scale and, then, extend them when the tests are successful. Development of food quality evaluation and process control systems is systems engineering. A system's engineering should be well scheduled and managed. Engineering is science, and it is also art. When solving a problem, researchers and developers need to incorporate the available techniques with the characteristics of the specific food object or process to discover unique solutions to the problem. In this book, we have tried to give a sense of science and art both for the development and integration of food quality evaluation and process control systems for food engineering automation. We hope our efforts lead you to success in your career.

References

Bourne, S. R., *The UNIX System*, Addison-Wesley, Reading, MA, 1982.

Huang, Y., Whittaker, A. D., Lacey, R. E., and Castell, M. E., Database information management through the World Wide Web for food science and engineering, ASAE/CSAE-SGCR Annual International Meeting, Toronto, Canada, 1999.

Ritchie, D. and Kernighan, B., *The C Programming Language*, Prentice-Hall, Englewood Cliffs, NJ, 1978.

Stroustrup, B., *The C++ Programming Language*, Addison-Wesley, Reading, MA, 1985.

Index